IDEAS from the
ARITHMETIC TEACHER
Grades 4–6   Intermediate School

compiled by

George Immerzeel

Bob Wills

from original IDEAS prepared by

George Bright

Marilyn Burns

Joan Duea

George Immerzeel

Earl Ockenga

Don Wiederanders

Copyright © 1979 by
The National Council of Teachers of Mathematics, Inc.
1906 Association Drive, Reston, Virginia 22091
All rights reserved
Third printing 1984
Printed in the United States of America
ISBN 0-87353-143-4

Introduction

The IDEAS section has been a feature of the *Arithmetic Teacher* since 1971. This collection has been selected from those activities appropriate for students in grades 4 through 8. The selections have been reprinted just as they originally appeared in the journal.

On one side of each page you will find the Pupil Activity Sheet; the teacher directions are on the reverse. This booklet has been perforated so that the pages can be easily removed and reproduced for classroom use. We suggest that you make a file of these pages or punch them for storage in a loose-leaf binder. Copies should be kept in the same file or binder so that you can use them when they are needed.

This volume has been topically arranged so that IDEAS for computational skills, for example, appear in one section, IDEAS for problem solving are grouped in another section, and so on. Suggested grade levels appear in the teacher directions for each IDEAS sheet.

Table of Contents

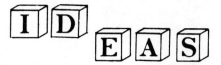

CONSECUTIVE NUMBER SUMS

Try to write all the numbers below as the sum of 2 or more consecutive numbers.

1 =

2 =

3 =

4 =

5 =

6 =

7 =

8 =

9 =

10 =

11 =

12 =

13 =

14 =

15 =

16 =

17 =

18 =

19 =

20 =

21 =

22 =

23 =

24 =

25 =

LOOK FOR PATTERNS!

For Teachers

Levels: 4, 5, 6, 7, 8

Objective: To investigate number patterns

Directions for teachers:

1. Duplicate a worksheet for each child.
2. Make sure they understand the directions (especially the meaning of *consecutive*).

Going further:

1. You may extend the activity to 50. It's good addition reinforcement as well as giving the students a chance to use the patterns they've found. Sums for all the numbers except the powers of 2 (1, 2, 4, 8, 16, . . .) are possible.
2. How many different ways are there to write each number as the sum of consecutive numbers? For example:

$$7 + 8 = 15$$
$$4 + 5 + 6 = 15$$
$$1 + 2 + 3 + 4 + 5 = 15$$

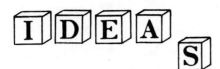

CROSS OUT SINGLES

Round 1

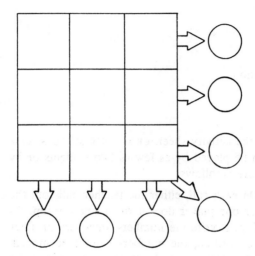

Score _____

Round 2

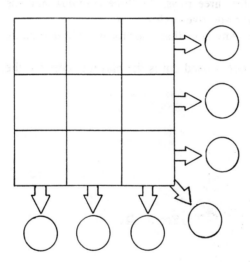

Score _____

Round 3

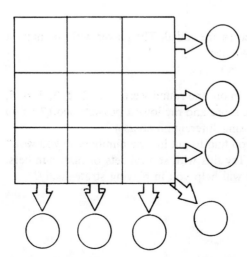

Score _____

GRAND TOTAL

For Teachers

Objective: Reinforcement in addition

Level: 2, 3, or 4

Directions:

Each student needs a copy of the activity sheet, CROSS OUT SINGLES. One die is also needed. The game can be played by as few as two students or by the entire class. The rules for play are as follows:

1. The die is rolled nine times. On each roll, either the teacher calls out the numbers for the entire class, or one player does it for a small group. As the numbers are called, the players write the numbers anywhere on their respective round-1 charts. Once written, the numbers cannot be moved.
2. Players then find the sums of the three rows, the three columns, and the diagonal, and record them in the respective circles.
3. Players check their own sums. Any sum that appears *in only one circle* must be crossed out.
4. The total of the sums that are *not* crossed out is the player's score for the round.

Example:

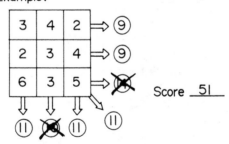

Score __51__

5. After three rounds, the three scores are added. The player with the highest total for the three rounds is the winner.

Going further:

1. Suppose the nine numbers thrown on one round were 1, 1, 2, 2, 3, 3, 4, 5, 6. What are the highest possible score and the lowest possible score? (You can change the problem by choosing different numbers.)
2. How many different ways can you find to put in nine numbers so you won't have any singles to cross out? Try this with several sets of nine numbers. Can you find any patterns that will help you in playing strategically?

Pick any number from 1 to 9 and write it in the box

START

Now, in your head, add 9 to the number in the box, and write the sum here. (Don't write the 9)

Continue adding 9 until you get to the double line

ANSWER THESE QUESTIONS:

1)   What's the pattern of the numbers in the 1s place? _____
_____

2)   What's the pattern of the numbers in the 10s and 100s places?_____
_____

3)   Add the digits of each number you wrote. What's the pattern?_____
_____

(If you add the digits and you get a 2-digit answer, add once more;
for example, 39. 3 + 9 = 12. 1 + 2 = 3.

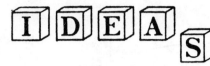 For Teachers

Level: 2, 3, 4, 5

Objective: To provide addition drill; to investigate number patterns

Directions for teachers:

1. Duplicate a worksheet for each child.
2. Make sure they understand the directions.

3. Have the children compare their results with others.

Going further:

1. Try the activity again adding 8s instead of 9s. Investigate the patterns.
2. Investigate what happens when adding other numbers also.

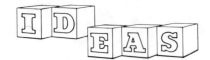

Name _____

You are to put the numbers from 1 to 9 in the circles of each triangle

Put them here so that
when you add up the
numbers on each side,
the three sums are all
the same and the
<u>smallest</u> possible sum.

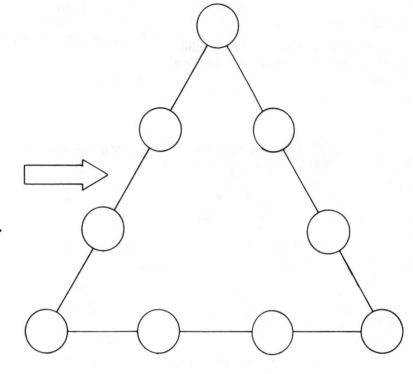

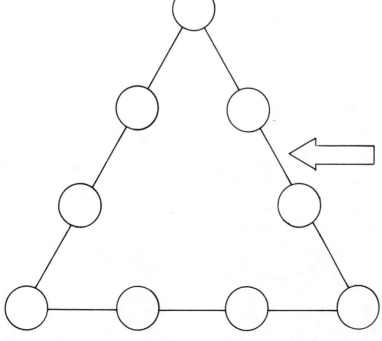

Put them here so that when
you add up the numbers
on each side, the three
sums are all the same and
the <u>largest</u> possible sum.

Level: 4, 5, 6, 7, 8

Objective: To provide drill in addition while solving a logical puzzle

Directions for teachers:

1. Duplicate a worksheet for each child.
2. Make sure they understand the directions.

For Teachers

Going further:

1. See if children can develop theories for how to solve the problem other than by trial and error.
2. Make up similar problems using other shapes and numbers of circles.

Answers:

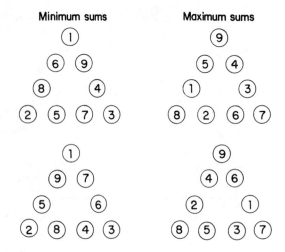

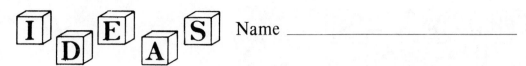

Name _____

R_1	1	2	3	4	5
R_2	10	9	8	7	6
R_3	11	12	13	14	15
R_4	20	19	18	17	16
R_5	21	22	23	24	25
R_6	30	29	28	27	26

1. If $R_1 + R_2$ means add all the numbers in row 1 and row 2, find these sums:
   a. $R_1 + R_2$ _____
   b. $R_2 + R_3$ _____
   c. $R_3 + R_4$ _____
   d. $R_5 + R_6$ _____
   e. $R_1 + R_2 + R_3 + R_4$ _____
   f. $R_1 + R_2 + R_3 + R_4 + R_5 + R_6$ _____

2. What is the sum of—
   a. The first 10 numbers? _____   c. The first 25 numbers? _____
   b. The first 20 numbers? _____   d. The first 30 numbers? _____

	1	2	3	4	5	6	7	8	9	10
R_1	1	2	3	4	5	6	7	8	9	10
R_2	20	19	18	17	16	15	14	13	12	11
R_3	21	22	23	24	25	26	27	28	29	30
R_4	40	39	38	37	36	35	34	33	32	31
R_5	41	42	43	44	45	46	47	48	49	50
R_6	60	59	58	57	56	55	54	53		
R_7	61	62	63	64	65					
R_8	80	79	78							
R_9	81	82	83							
R_{10}	100	99								

3. Find the sum of—
   a. $R_1 + R_2$ _____
   b. $R_3 + R_4$ _____
   c. $R_5 + R_6$ _____
   d. $R_7 + R_8$ _____
   e. $R_9 + R_{10}$ _____
   f. The first 100 numbers _____

 For Teachers

Objective: Experience with patterns in solving the sums of basic arithmetic series

Grade level: 5, 6, 7, or 8

Directions for teachers:

1. Remove the student worksheet and reproduce one copy for each student.
2. Have students do part or all of the exercises independently; then discuss their methods of solution.
3. Be sure the student understands that though some numbers are not pictured, all rows in the larger table contain ten numbers.

Comments: The computation in this lesson will be either very tedious or relatively simple, depending on whether or not the student makes use of the patterns in the tables. The teacher will be tempted to tell the students how to avoid the tedious task; by doing so, however, the teacher denies the student the opportunity to learn from his own experience. The student who observes that $R_1 + R_2$ can be found in the first table by taking 5×11 and in the second table by taking 10×21 is well on his way toward appreciating the role of patterns in mathematics. This appreciation is acquired most surely by the student who has personally struggled with tasks that are made easy when he discovers the pattern.

These activities could be expanded to include the investigation of sums of sets of even numbers, sums of sets of odd numbers, or sums of sets of multiples of whole numbers—such as $10 + 20 + 30 + \ldots + 100$. This last example provides excellent experience with the distributive property—that is, $10 + 20 + 30 + 40 = 10 (1 + 2 + 3 + 4)$, and so forth.

Answers

1.   a. 55   b. 105   c. 155   d. 255   e. 210   f. 465

2.   a. 55   b. 210   c. 325   d. 465

3.   a. 210   b. 610   c. 1010   d. 1410   e. 1810   f. 5050

Match the number tags to the spaces. Use each tag only once.

A

$-\ 2$

B

$-\ 7$

C

$-\ 4$

D $\quad - 5 =$

E $\quad - 1 =$

Cut out the tags.

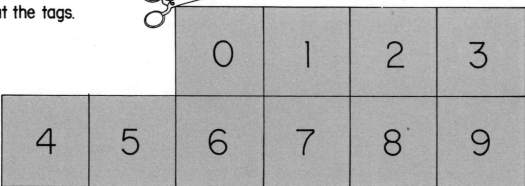

| 0 | 1 | 2 | 3 |
| 4 | 5 | 6 | 7 | 8 | 9 |

For Teachers

Objective: Experience with subtraction facts

Levels: 2, 3, 4

Directions for teachers:

1. Remove the student activity sheet and reproduce one copy for each child.
2. Encourage students to try various arrangements of their number tags so each tag is used once and all five examples are correct.

Comments:

After the students have tried the activity alone, modifications could be introduced:

The students who have not completed all five subtraction examples could work together in pairs.

Students who have been successful might be challenged to complete the examples with only five tags. In this activity, the tags could be used more than once. A scoring system could also be introduced—one point is awarded for every correctly completed example.

Answers:

Students have found six solutions. The following is one.

A	8	B	7	C	5	D	9 − 5 = 4
	− 2		− 7		− 4		
	6		0		1	E	3 − 1 = 2

IDEAS

Mark each play to
find the winner.

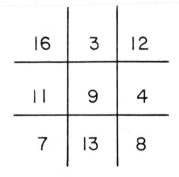

8	12	13
7	11	9
6	10	14

Plays

$\square$ = 6 + 5 $\square$ = 2 + 4

$\bigcirc$ = 8 + 6 $\bigcirc$ = 7 + 6

$\square$ = 5 + 3 $\square$ = 4 + 3

$\bigcirc$ = 8 + 2 $\bigcirc$ = 7 + 5

16	3	12
11	9	4
7	13	8

Plays

$\square$ = 9 + 7 $\square$ = 10 − 6

$\bigcirc$ = 11 − 3 $\bigcirc$ = 7 + 6

$\square$ = 12 − 9 $\square$ = 6 + 5

$\bigcirc$ = 7 + 5 $\bigcirc$ = 15 − 8

11	8	9
4	6	5
15	17	14

Plays

3 + $\bigcirc$ = 12   7 + $\bigcirc$ = 15

14 − 8 = $\square$ $\square$ = 3 + 8

6 + 9 = $\bigcirc$ 9 + $\bigcirc$ = 14

$\square$ + 4 = 11   6 + 8 = $\square$

 For Teachers

Objective: Experience with the basic facts of addition and subtraction

Levels: 3 or 4

Directions for teachers:

1. Remove the activity sheet and reproduce a copy for each student.
2. Study the example. Note that one player marks with ☐ and the other with ○.
3. Make sure the directions are understood. (That is, the first player puts a ☐ around 11, the second player puts a ○ around 14, and so on. Three like marks in a row is a winner.)

Comments: Note that the game is a take-off on the familiar tic-tac-toe. A similar experience with multiplication facts and missing factor equations (3 × ☐ = 21, and so on) may be appropriate for your students. Supplemental exercises are easily created if you "play the game" as you list the equations. The fact that it is difficult to "win" doesn't seem to affect the students' enthusiasm for playing the game.

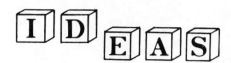

Name _____

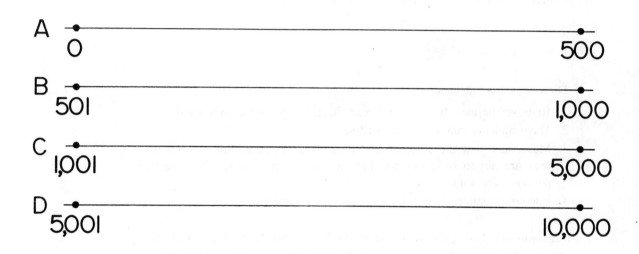

A   0 ●————————————————————● 500

B   501 ●————————————————————● 1,000

C   1,001 ●————————————————————● 5,000

D   5,001 ●————————————————————● 10,000

Which number line has a point for these numbers?

_____ 247	_____ 965	_____ 1245
_____ 4,645	_____ 9,999	_____ 4965
_____ 8,492	_____ 873	_____ 1468
_____ 200 + 840	_____ 654 + 87	_____ 982 + 1025
_____ 146 + 283	_____ 845 + 4525	_____ 96 + 1025
_____ 999 + 3	_____ 2042 + 4650	_____ 921 + 64

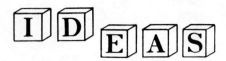

For Teachers

Objective: Experience in estimating sums.

Levels: 4 or 5

Directions for teachers:

1. Remove the activity sheet and reproduce a copy for each student.
2. Have students study each number line.
3. Make sure that the students understand that when addition is involved they are not to write the sum but only to identify the number line that contains the sum.
4. Encourage students to respond without computing the sums.

Comments: Estimation is a seldom-taught, but highly desirable, skill. The cautious student will carefully compute each sum before deciding on his response. Students should be couraged to be risk takers. You can readily expand on this experience. Easier sums fall near the center of the segments. In more challenging exercises, the sums are "close to" the identified points.

Clues to answers: 2 A's, 4 B's, 8 C's, 4 D's.

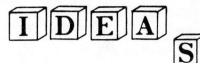

One player spins this spinner.

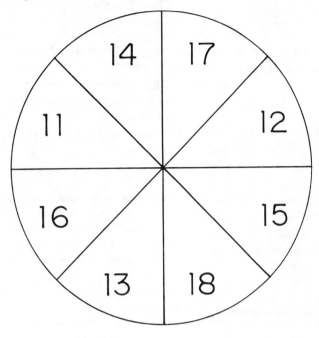

The other player spins this spinner.

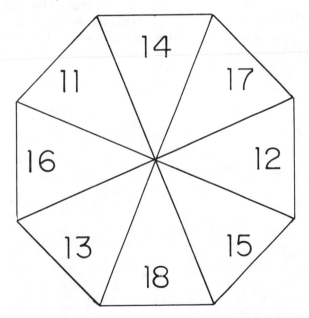

Player with the larger number scores as many points as the difference.

Winner is the first player to get 15 points.

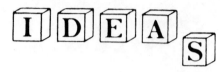

Objective: Experience with subtraction, with a focus on difference

Levels: 2, 3, 4

Directions for teachers:

1. Remove the student activity sheet and reproduce one copy for each group of students.

2. To make the spinners, place Scotch tape over the center of the spinner circles. Push a thumbtack from the back of the paper through the center of the circle. Attach a short (6 to 8 cm) pencil to the point of the thumbtack.

3. This activity can be played by two students per group.

Directions for pairs of students:

1. One player spins the circle spinner. The other player spins the octagon spinner.

2. Players decide who has spun the larger number.

3. The difference between the two numbers is the score of the player with the larger number.

4. Winner is the first player to get at least 15 points.

Comments:

If students have had minimal experiences in comparing sets to determine differences, objects such as lima beans or paper clips could be used to provide a physical referent. By using one-to-one correspondence, they can readily see which player has the larger number and determine the difference (how many more).

Students might make the activity more challenging by changing the numbers on the spinners or changing the scoring of this activity. One student suggested playing odd and even numbers. One player scored a point when the difference was an even number and the other player scored a point when the difference was an odd number.

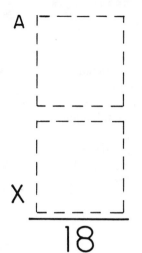

Name_____

Match the number tags to the spaces. Use each tag only once.

A
┌─────┐
│   │
│   │
└─────┘
┌─────┐
X │   │
└─────┘
─────────
    18

B
┌─────┐
│   │
│   │
└─────┘
┌─────┐
X │   │
└─────┘
─────────
    24

C
┌─────┐
│   │
│   │
└─────┘
┌─────┐
X │   │
└─────┘
─────────
    40

D
┌─────┐   ┌─────┐
│   │ X   │   │   = 0
└─────┘   └─────┘

E
┌─────┐   ┌─────┐
│   │ X   │   │   =21
└─────┘   └─────┘

Cut out the tags.

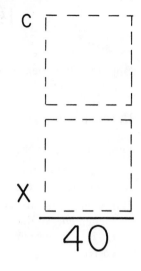

0	1	2	3

4	5	6	7	8	9

 For Teachers

Objective: Experience with multiplication facts

Levels: 4, 5, 6

Directions for teachers:

1. Remove the student activity sheet and reproduce one copy for each child.
2. Encourage students to try various arrangements of their number tags so each tag is used once and all five examples are correct.

Comments:

After the students have tried this activity alone, modifications could be introduced:

The students who have not completed all five multiplication problems could work together in pairs.

Students who have been successful might work in pairs and take turns drawing five tags. These tags could be used more than once, with students receiving one point for each multiplication problem correctly completed.

Answers:

A 9   B 6   C 8   D 1 x 0 = 0
  x 2   x 4   x 5
  ‾‾18‾   ‾‾24‾   ‾‾40‾   E 7 x 3 = 21

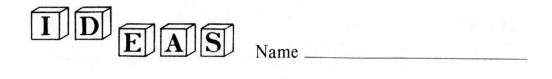

Name _____

Directions: Cut out the squares. Fit them together so that the edges that touch name the same number.

27	24		54
9×8 24	40 9×5	10×10 36 72	
6×8	7×3	7×6	6×4
49	48		56
81	45 64	8	25
	48	6×5	
63	8×6	21	20
8×8	70 5×5	9×9 10×7	5×8
8×7			7×7
42		30	
6×6	100 8×0	8×3	0 8×1
5×4	9×6	7×9	9×3

IDEAS   For Teachers

for Levels 4, 5, or 6

Objective: Experience with the concept of equals.

Directions:
1. Provide each student with a copy of the appropriate activity sheet and a pair of scissors.
2. If necessary, give the hint that when the squares are fit together correctly, a 4 by 4 square is once again found.

Comments: The fewer directions required the better. Students need experience in figuring things out for themselves. As the student progresses with each puzzle, his decisions receive immediate reinforcement.

Key:

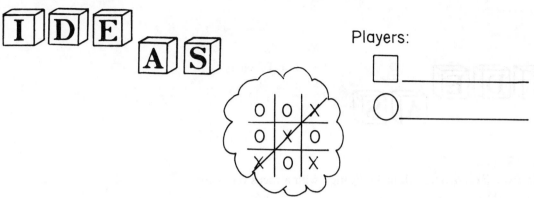

Rules: 1. Play by crossing out a number in your set. (Players take turns.)

2. When the <u>product</u> of your "play" and the <u>last play</u> of your opponent is on the grid (#), mark it with your mark (□ or ○).

3. Win by getting three of your marks in a row.

Game 1

12	25	56
36	15	32
42	0	24

□ plays { 0, 1, 2, 3, 4, 5, 6, 7, 8, 9 }

○ plays { 0, 1, 2, 3, 4, 5, 6, 7, 8, 9 }

Game 2

27	30	24
36	18	12
32	48	25

○ plays { 0, 1, 2, 3, 4, 5, 6, 7, 8, 9 }

□ plays { 0, 1, 2, 3, 4, 5, 6, 7, 8, 9 }

Game 3

24	30	27
40	36	42
45	16	18

□ plays { 0, 1, 2, 3, 4, 5, 6, 7, 8, 9 }

○ plays { 0, 1, 2, 3, 4, 5, 6, 7, 8, 9 }

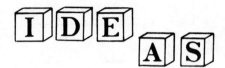

Objective: Experience with multiplication with a focus on factors.

Levels: 4, 5, or 6

Directions for teachers:

1. Remove the activity sheet and reproduce a copy for each student.
2. Play the first game with the class, you against the class. Copy the grid and sets of factors on the board. Here is a possible play for Game 1:

Turns	1	2	3	4	5	6	7	8	9	10	
□	2	7	3	0	4	6	8	1	9	5	Win
○	6	8	0	9	7	4	2	1	5		

3. Have students play games 2 and 3 in pairs.
4. Students can make up their own games by taking turns making the game grids.

Comments: This take-off on tic-tac-toe involves good practice with the multiplication basic facts. The focus on the factors is especially beneficial to many students. All students can feel some level of success. The game strategist who can plan several plays ahead has a distinct advantage. A skillful player can force his opponent to let him win.

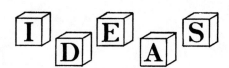

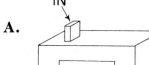

A.

Study the table. Complete the entries.

IN	3	5	7	9	10	12
OUT	9	15				

B.

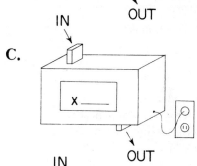

IN	4	7	2	9	5	8
OUT	20	35				

C.

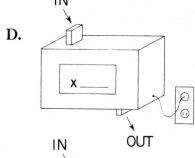

IN	7	3	8	5	10	
OUT	70	30				150

D.

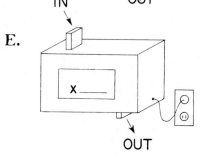

IN	3	6	10	4	7	
OUT	27	54				45

E.

IN	5	2	4		10	
OUT	35	14		28		42

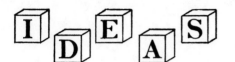 For Teachers

Objective: Experience with patterns in factors and products

Grade level: 4 or 5

Directions for teachers:

1. Remove the activity sheet and reproduce a copy for each student.
2. Study and complete the table for machine *A* as a class.
3. Discuss what machine *A* is programmed to do. *Answer:* multiply by 3.
4. Have students complete the tables for machines *B, C, D,* and *E*.
5. Discuss each machine.

Comments: The temptation to have a more perceptive student tell everyone what to do for each machine should be avoided. The slower student may need more clues (entries in the table) before he sees the pattern. Giving him more clues is far better for the experience with patterns than telling him what the rule is. The latter may reduce the activity to a drill on the basic facts. Variations on these machines and tables will provide individualized experiences for a wide range of student ability.

Answers

Machine *B:* multiply by 5; Machine *C:* multiply by 10; Machine *D:* multiply by 9; Machine *E:* multiply by 7.

Find the sum of the numbers in each table.

A.

2 x 3	2 x 5	2 x 2	
4 x 3	4 x 5	4 x 2	
5 x 3	5 x 5	5 x 2	

Sum

B.

3 x 4	3 x 5	3 x 6	
5 x 4	5 x 5	5 x 6	
6 x 4	6 x 5	6 x 6	

Sum

C.

2 x 1	2 x 3	2 x 6	
3 x 1	3 x 3	3 x 6	
5 x 1	5 x 3	5 x 6	

Sum

D.

1 x 5	1 x 7	1 x 8	
2 x 5	2 x 7	2 x 8	
3 x 5	3 x 7	3 x 8	

Sum

For Teachers

Objective: Experiences with patterns that focus on the distributive property

Grade level: 4, 5, or 6

Directions for teachers:

1. Reproduce a copy of the worksheet for each student.

2. Have the students find the sum for each table.

3. When the faster students have finished, copy table A on the chalkboard and have students tell how they found the sum. Draw out as many different techniques as possible.

4. If no one brings up a shortcut involving the distributive concept, make another copy of the table on the chalkboard and ask for help in completing your table. Note that the sum 110 could be found from the bottom row (11 × 10) or the right column (11 × 10).

2 x 3	2 x 5	2 x 2	2 x ?
4 x 3	4 x 5	4 x 2	4 x ?
5 x 3	5 x 5	5 x 2	5 x ?
? x 3	? x 5	? x 2	

Sum

5. Have students discuss the patterns in each of the other tables.

Comments: The distributive property is the most elusive of the basic properties for the learner. The experiences involved in this worksheet will give some students an opportunity to identify the distributive pattern and see how it provides for efficient computation. Some students will have had only a pattern-oriented experience involving application of the multiplication basic facts and some addition. Those students that have had previous exposure to the distributive property will recognize its impact on this task. Those students that have not been formally introduced to the distributive property will more readily identify it when it is encountered.

Answers

A. 110   B. 210   C. 100   D. 120

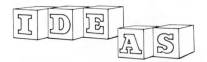

Take the number 7. If you had 7 squares of paper, there is only 1 way to make a rectangular shape

That's why 7 is colored on the chart

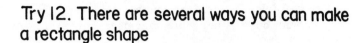

Try 12. There are several ways you can make a rectangle shape

Too many rectangles -- so 12 is not colored in

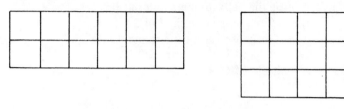

Investigate all the numbers on the chart and color those for which there is only 1 way to make a rectangle shape

1	2	3	4	5	6
7	8	9	10	11	12
13	14	15	16	17	18
19	20	21	22	23	24
25	26	27	28	29	30
31	32	33	34	35	36

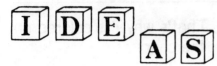

Level: 4, 5, 6, 7, 8

Objective: To investigate prime numbers

Directions for teachers:

1. Duplicate a worksheet for each student.
2. Make sure they understand the directions.

Going further:

1. Discuss prime numbers with your

class—all the prime numbers will be colored on the completed worksheets. A prime number has 2 distinct factors, itself and 1. Does the number 1 have two distinct factors? The number 1 is colored, but 1 is not a prime number.

2. Notice that all numbers colored below the first row appear in the 1 or 5 column. That's because the prime numbers, excepting 2 and 3, are 1 more or 1 less than a multiple of 6.

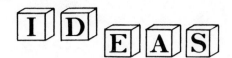

Put the letters in the right box.

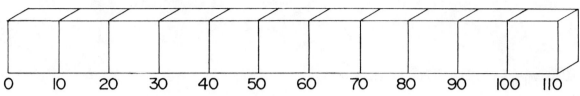

R	9×7	T	21×5
L	6×3	R	25×3
E	9×9	C	7×7
O	6×9	L	12×2
A	4×2	C	27×3

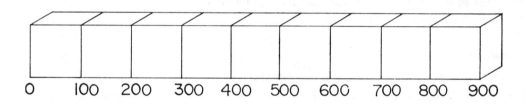

N	37×21	L	15×23
O	37×17	E	43×22
L	26×9	W	13×7
E	13×13	D	29×18

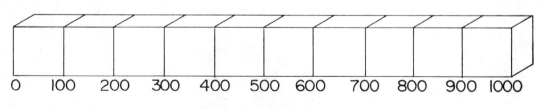

R	29×29	W	25×25
I	16×16	E	21×21
N	19×19	F	12×12
O	27×27	K	31×31

IDEAS   For Teachers

IDEAS   For Levels 4 or 5

Objective: Experiences in ordering numbers that encourage estimation rather
than computation using a modification of the number line

Directions for teachers:

1. Give each student a copy of the appropriate activity sheet.
2. Have them read the directions and go to work.
3. Observe each student individually to be sure that the directions are
   carried out.

Comments: After most of the students have completed naming the *second
row* of boxes, you may wish to encourage estimating by commenting: "Some
students seem to be able to figure out which box a letter goes on by estimating
rather than doing the computation."

1	2	3	4	5	6	7	8	9	10	11	12
2	4	6	8	10	12	14	16	18	20	22	24
3	6	9	13	5	18	21	24	27	30	33	36
4	8	12	16	20	24	28	32	36	42	44	48
5	10	15	20	25	30	35	40	45	50	55	60
6	12	18	24	30	36	42	48	54	60	66	72
7	14	21	28	35	42	49	56	63	70	77	84
8	16	24	32	40	48	56	64	72	80	88	96
9	18	27	36	45	54	63	72	81	90	99	108
10	20	30	40	50	60	70	80	90	100	110	120
11	22	33	44	55	66	77	88	99	110	121	132
12	24	36	48	60	72	84	96	108	120	132	144

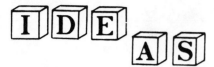

For Teachers

Levels: 4, 5, 6

Objective: To give practice with multiples and the multiplication chart

Directions for teachers:

1. Duplicate a worksheet for each student.
2. Any number from 2 to 12 may go in the box. The student copies the multiples of that number from the chart and then continues the multiples down to a multiple closest to but no greater than 144.
3. Then the student colors every number on the list wherever it appears on the chart. It helps to have children cross the numbers off the list as they color the chart.
4. The charts should be posted and the patterns for the different numbers compared.

Going further:

If children wish to make their own complete "Multiplication Chart Pattern Books," they need to do 11 sheets, one for each of the numbers from 2 to 12. It is an ambitious project, but may interest some students.

Sample:

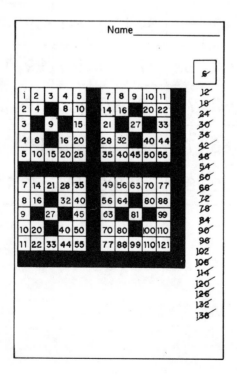

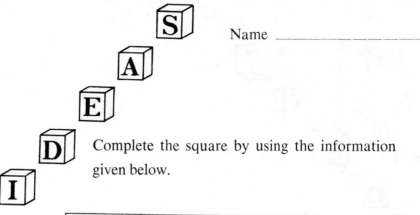

Name _____

Complete the square by using the information given below.

a	b	c	d
e	f	g	h
i	j	k	l
m	n	o	p

Given:

a. 9×20

b. 24×5

c. 12×3

d. 9×8

e. 6×8

f. 12×5

g. 24×8

h. 9×12

i. $6 \times 7 \times 4$

j. $3 \times 11 \times 4$

k. 6×4

l. $7 \times 3 \times 4$

m. 3×4

n. $6 \times 8 \times 2$

o. $13 \times 6 \times 2$

p. $3 \times 12 \times 4$

Check your answers by adding in any direction. The totals should be the same.

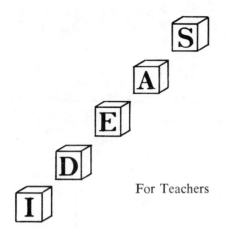

For Teachers

Objective: Computation practice with multiplication.

Grade level: 4, 5, or 6

Directions for teacher:

1. Tear out student worksheet and reproduce one copy for each student.

Directions to be read to students:

1. What is the standard name for 6 × 5? for 9 × 8? On this worksheet write the standard name for the numbers given at the bottom of the sheet in the squares at the top.

2. When you have finished add the rows or the columns or the diagonals.

Comments: A similar worksheet can be made from any magic square by multiplying the entries by any factor you choose. Not only can this kind of exercise provide practice in an interesting manner, but the magic square also gives immediate reinforcement for correct answers.

Other examples of magic square are:

4	9	2
3	5	7
8	1	6

7	12	1	14
2	13	8	11
16	3	10	5
9	6	15	4

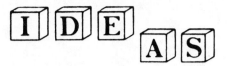

Name _____

Which number line has a point for these numbers?

_____ 1. 9 x 8   _____ 2. 25 x 12   _____ 3. 51 x 12

_____ 4. 21 x 40   _____ 5. 45 x 33   _____ 6. 18 x 72

_____ 7. 92 x 9   _____ 8. 25 x 25   _____ 9. 323 x 5

_____ 10. 35 x 16   _____ 11. 105 x 12   _____ 12. 14 x 8

_____ 13. 32 x 61   _____ 14. 87 x 8   _____ 15. 54 x 23

To help you check: There are 3 points on line A, 6 points on
line B, 4 points on line C, and 2 points on line D.

For Teachers

Objective: Experience in estimating products

Levels: 6, 7, or 8

Directions for teachers:

1. Remove the activity sheet and reproduce a copy for each student.
2. Have the students personally study each number line.
3. Make sure that the students understand that they are to simply identify the number line that contains the product. They should not compute the products.
4. Encourage students to respond without computing the products.

Comments: You will have to be very forceful if you really wish to discourage some students from computing the products before they respond. Many students are basically not risk takers. Some are very insecure in the process of estimation. You can readily expand on this experience. In easier exercises, the points can be near the center of the segments. More challenging exercises can have products that are "close to" the identified points.

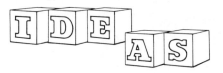

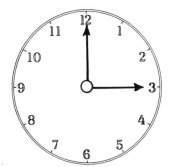

This clock is set at 3:00

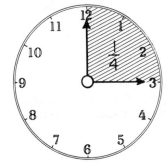

If it's shaded this way, it shows $\frac{1}{4}$

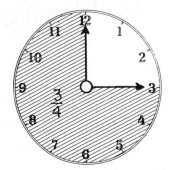

If it's shaded this way, it shows $\frac{3}{4}$

For each clock face: 1. Draw the hands for the time shown.
2. Shade in one section.
3. Write a fraction that tells how much you shaded.

9:00
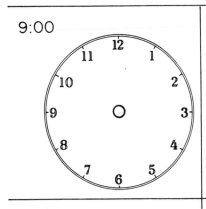

2:00
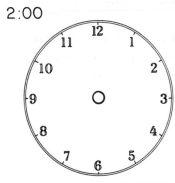

8:00
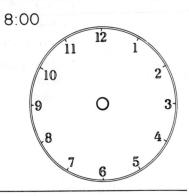

1:00
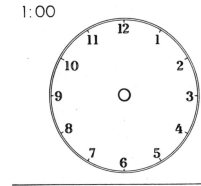

6:00
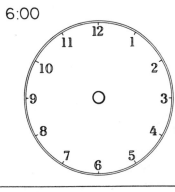

4:00
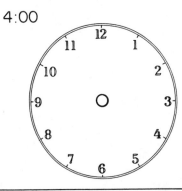

11:00
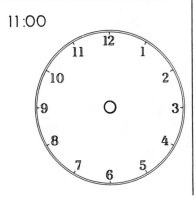

10:00
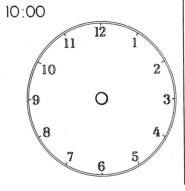

5:00
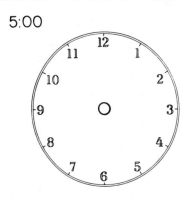

 For Teachers

Objective: To identify fractions with the position of the hands on a clock

Levels: 4 and 5

Directions for teachers:

The worksheets suggest ways to use clockfaces for experiences with fractions as well as practice in telling time.

1. Give each student a copy of the worksheet.

2. Go over the directions with the students.

3. Discuss the sheet, emphasizing the two fractions that are possible for each clock.

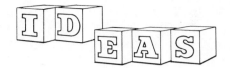

Circle one fraction.

$$\frac{1}{4} \qquad \frac{1}{3} \qquad \frac{3}{4} \qquad \frac{1}{6} \qquad \frac{5}{6} \qquad \frac{1}{12} \qquad \frac{5}{12}$$

On each clock, draw hands so you can shade in part to match the fraction you chose. Use the same fraction for all clocks. Under each clock, write the time that most closely matches the hands.

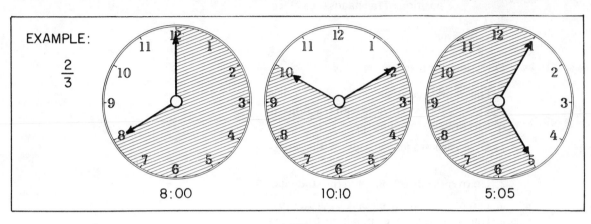

EXAMPLE:

$\frac{2}{3}$

8:00   10:10   5:05

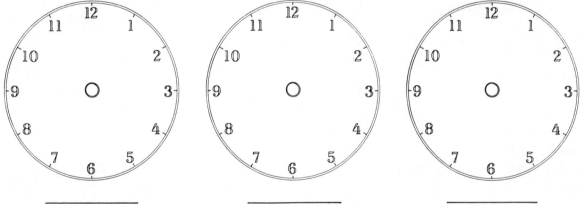

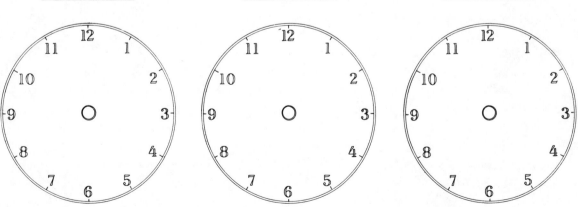

Use another worksheet if you want to try a different fraction.

 For Teachers

Objective: To identify fractions with the position of the hands on a clock

Levels: 5 and 6

Directions for teachers:

The worksheets suggest ways to use clockfaces for experiences with fractions as well as practice in telling time.

1. Give each student a copy of the worksheet.

2. Go over the directions with the students.

3. Discuss the sheet, emphasizing the two fractions that are possible for each clock.

For each time, draw the hands and shade in the smaller section of the clock. Find a fraction name for the section of the clock that is shaded.

EXAMPLE: 3:43

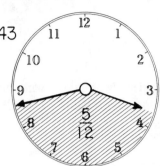

$\frac{5}{12}$

2:05	8:05	5:50
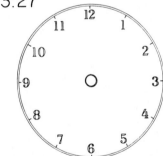		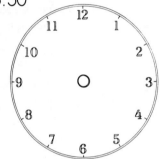

3:27	10:16	6:05
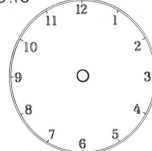		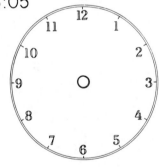

Make up your own. Exchange with another student and check each other's problems.

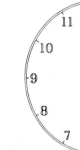

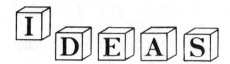 For Teachers

Objective: To identify fractions with the
position of the hands on a clock

Levels: 5 and 6

Directions for teachers:

The worksheets suggest ways to use
clockfaces for experiences with fractions as
well as practice in telling time.

1. Give each student a copy of the worksheet.

2. Go over the directions with the students.

3. Discuss the sheet, emphasizing the two
   fractions that are possible for each
   clock.

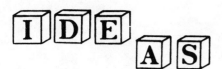

Name _____

Use a ruler to draw a line segment between each pair of equivalent fractions.

Color the three squares.

$\frac{3}{4}$•

•$\frac{2}{6}$

•$\frac{1}{2}$

$\frac{1}{4}$•

•$\frac{2}{8}$

$\frac{2}{10}$•

•$\frac{1}{5}$

$\frac{6}{8}$

•$\frac{2}{3}$

$\frac{90}{100}$•

•$\frac{9}{10}$

$\frac{3}{5}$•

•$\frac{6}{10}$

•$\frac{5}{10}$ •$\frac{6}{9}$

•$\frac{1}{3}$

 For Teachers

Objective: To practice identifying equivalent fractions

Levels: 3, 4, 5

Directions for teachers:

1. Give each student a copy of the worksheet.
2. Read the directions with the students. The students will draw more accurate figures if they use a ruler or straightedge. The squares will be recognizable, however, if a straightedge is not used.
3. You may want to give an example of equivalent fractions before the students begin.

Answer:

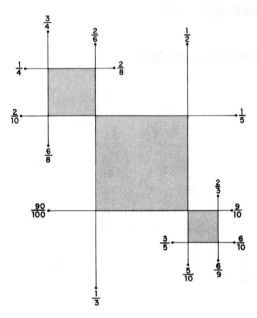

Use a ruler to draw a line segment between each pair of fractions whose sum is 1.

Count all the squares

$\frac{2}{3}$•   •$\frac{2}{8}$

$\frac{1}{2}$•   •$\frac{11}{12}$

$\frac{7}{12}$•   •$\frac{7}{10}$

$\frac{4}{5}$•   •$\frac{5}{8}$

$\frac{3}{4}$•   •$\frac{1}{3}$

$\frac{1}{12}$•   •$\frac{2}{4}$

$\frac{3}{10}$•   •$\frac{5}{12}$

$\frac{3}{8}$•   •$\frac{2}{10}$

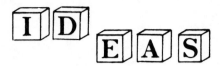

For Teachers

Objective: To practice adding fractions whose sum is 1

Levels: 5, 6, 7

Directions for teachers:

1. Give each student a copy of the worksheet.

2. Let the students read the directions. Emphasize that the students are looking for pairs of fractions whose sum is 1; for example, $\frac{3}{8} + \frac{5}{8}$ or $\frac{2}{3} + \frac{1}{3}$. The use of straightedges will make the drawings more accurate, but you do not need to insist on their use.

3. Three fractions, $\frac{2}{8}$, $\frac{2}{4}$, and $\frac{2}{10}$, are not in lowest terms. If students have trouble matching these fractions with

   $$\frac{3}{4}, \frac{1}{2}, \text{ and } \frac{4}{5},$$

   respectively, suggest to the students that they rewrite all fractions in lowest terms before trying to match them.

4. After the students have drawn all the line segments, encourage them to identify not only the simplest squares; for example, ◇ ; but also more complex ones.

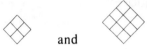

and

The total number of squares is 14.

Answer:

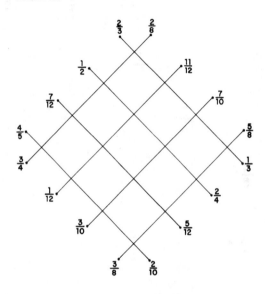

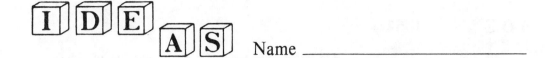

Name _____

Directions: Cut out the squares. Fit them together so that the edges that touch name the same number.

	$\frac{2}{4}$		2		$\frac{7}{10}$		
1		$\frac{4}{12}$	$\frac{2}{10}$		$\frac{1}{8}$	$\frac{3}{7}$	$\frac{2}{3}$
	$\frac{1}{7}$			$\frac{3}{2}$			$\frac{1}{4}$
			$\frac{5}{3}$		$\frac{2}{14}$		$\frac{1}{3}$
$\frac{2}{12}$		$\frac{8}{10}$		$\frac{3}{3}$	$\frac{2}{16}$	$\frac{5}{8}$	$\frac{1}{2}$
	$\frac{3}{5}$		$\frac{6}{3}$		$\frac{3}{4}$		
	$\frac{6}{16}$		$\frac{6}{8}$		$\frac{10}{12}$		$1\frac{1}{2}$
$\frac{10}{16}$		$\frac{3}{10}$		$\frac{12}{24}$		$\frac{1}{5}$	$\frac{6}{14}$
	$\frac{2}{6}$				$.7$		
			$\frac{2}{8}$				$\frac{6}{10}$
$\frac{4}{6}$		$\frac{0}{2}$		$\frac{4}{5}$	0	$\frac{1}{6}$	$\frac{1}{3}$
	$\frac{2}{3}$		$\frac{5}{6}$		$\frac{1}{2}$		$\frac{3}{8}$

$.3$

Objective: Experience with the concept of equals.

Directions:
1. Provide each student with a copy of the appropriate activity sheet and a pair of scissors.
2. If necessary, give the hint that when the squares are fit together correctly, a 4 by 4 square is once again found.

Comments: The fewer directions required the better. Students need experience in figuring things out for themselves. As the student progresses with each puzzle, his decisions receive immediate reinforcement.

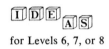

for Levels 6, 7, or 8

Key:

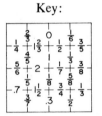

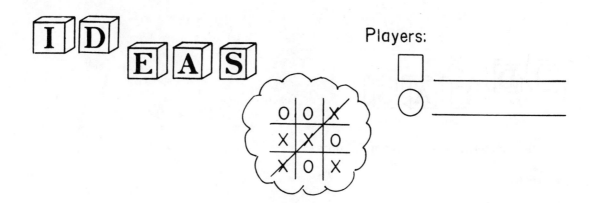

Players:

☐ _____

◯ _____

Rules: 1. Play by placing your mark (☐ or ◯) on a fraction.

2. Win **only** if you get three in a row **that are in order** (smallest to largest or largest to smallest).

$\frac{2}{5}$	$\frac{7}{10}$	$\frac{7}{8}$
$\frac{2}{3}$	$\frac{1}{2}$	$\frac{6}{10}$
$\frac{3}{4}$	$\frac{5}{8}$	$\frac{1}{8}$

Game 1

.9	.82	.7
.2	.62	.87
.35	.4	.5

Game 2

$\frac{3}{7}$	$\frac{2}{3}$	$\frac{1}{2}$
$\frac{3}{4}$	$\frac{4}{7}$	$\frac{4}{8}$
$\frac{1}{3}$	$\frac{1}{2}$	$\frac{5}{7}$

Game 3

.2	$\frac{1}{8}$	.125
$\frac{1}{3}$	$\frac{2}{5}$	.7
$\frac{1}{2}$	.6	$\frac{3}{4}$

Game 4

$\frac{3}{4}$	$\frac{2}{5}$	.75
.7	$\frac{3}{5}$	$\frac{1}{2}$
$\frac{2}{3}$	$\frac{4}{5}$	$\frac{3}{8}$

Game 5

.7	.65	.25
.52	.6	.75
.300	.2	.125

Game 6

 For Teachers

Objective: Experience with ordering fractions

Levels: 6, 7, or 8

Directions for teachers:

1. Remove the activity sheet and reproduce a copy for each student.
2. Have students read and discuss the directions. Be sure they understand that *two* criteria must be met to win.
3. After playing several games, have students discuss possible game strategies.
4. Have students try to make up game grids with as many potential ways to win as they can.

Comments: This take-off on tic-tac-toe forces the student to "read" the fraction symbols. You may wish to design more grids that have either common fractions only or decimal fractions only in them. Grids that are very easy or that are very hard may be constructed.

Name_____

Put the letters on the right box.

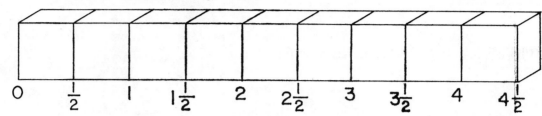

$$0 \quad \frac{1}{2} \quad 1 \quad 1\frac{1}{2} \quad 2 \quad 2\frac{1}{2} \quad 3 \quad 3\frac{1}{2} \quad 4 \quad 4\frac{1}{2}$$

L	$\frac{5}{8}$	*G*	$\frac{10}{3}$
I	$2\frac{2}{3}$	*A*	$\frac{3}{7}$
L	$\frac{10}{7}$	*T*	$4\frac{9}{20}$
H	$2\frac{5}{3}$	*R*	$\frac{9}{4}$

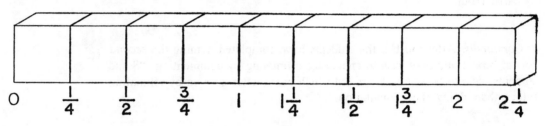

$$0 \quad \frac{1}{4} \quad \frac{1}{2} \quad \frac{3}{4} \quad 1 \quad 1\frac{1}{4} \quad 1\frac{1}{2} \quad 1\frac{3}{4} \quad 2 \quad 2\frac{1}{4}$$

D	$2\frac{1}{8}$	*Y*	$\frac{7}{8}$
O	$1\frac{2}{3}$	*O*	$\frac{15}{8}$
E	$\frac{3}{8}$	*V*	$1\frac{1}{5}$
R	$\frac{7}{10}$	*G*	$1\frac{3}{8}$

Which box does each go in ?

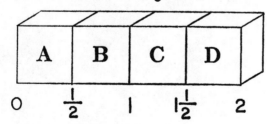

$$0 \quad \frac{1}{2} \quad 1 \quad 1\frac{1}{2} \quad 2$$

1. $\frac{3}{4}+\frac{1}{2}$ ____   4. $5\times\frac{1}{4}$ ____

2. $2\times\frac{1}{3}$ ____   5. $2-1\frac{1}{4}$ ____

3. $1\frac{1}{3}+\frac{1}{2}$ ____   6. $5-4\frac{1}{8}$ ____

IDEAS

IDEAS

For Levels 5 or 6

Objective: Experiences in ordering numbers that encourage estimation rather than computation using a modification of the number line

Directions for teachers:

1. Give each student a copy of the appropriate activity sheet.
2. Have them read the directions and go to work.
3. Observe each student individually to be sure that the directions are carried out.

Comments: After most of the students have completed naming the *second row* of boxes, you may wish to encourage estimating by commenting: "Some students seem to be able to figure out which box a letter goes on by estimating rather than doing the computation."

Name _____

Put the letters on the right box.

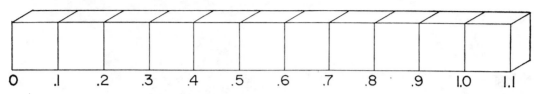

E	.15	*O*	.69
D	.87	*Y*	.399
O	.762	*G*	.55
R	.201	*V*	.03

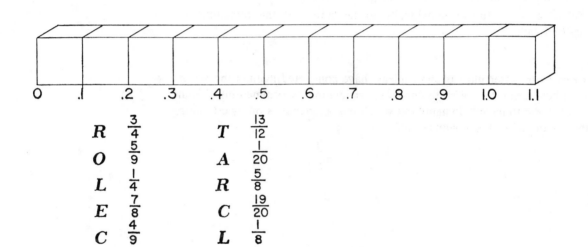

R	$\frac{3}{4}$	*T*	$\frac{13}{12}$
O	$\frac{5}{9}$	*A*	$\frac{1}{20}$
L	$\frac{1}{4}$	*R*	$\frac{5}{8}$
E	$\frac{7}{8}$	*C*	$\frac{19}{20}$
C	$\frac{4}{9}$	*L*	$\frac{1}{8}$

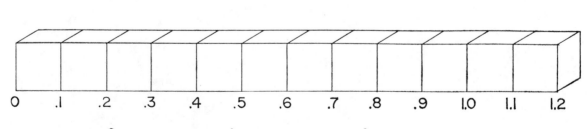

N	$\frac{2}{3}$	*H*	$\frac{1}{3}$	*R*	$1\frac{1}{9}$
E	$\frac{19}{20}$	*S*	$1\frac{1}{6}$	*T*	$\frac{5}{11}$
I	$\frac{17}{100}$	*G*	$\frac{1}{4}$	*A*	$\frac{13}{24}$
S	$\frac{3}{4}$	*W*	$\frac{5}{6}$	*R*	$\frac{1}{13}$

IDEAS

⬜️D️E️A️Ⓢ   For Levels 6, 7, or 8

Objective: Experiences in ordering numbers that encourage estimation rather than computation using a modification of the number line

Directions for teachers:

1. Give each student a copy of the appropriate activity sheet.
2. Have them read the directions and go to work.
3. Observe each student individually to be sure that the directions are carried out.

Comments: After most of the students have completed naming the *second row* of boxes, you may wish to encourage estimating by commenting: "Some students seem to be able to figure out which box a letter goes on by estimating rather than doing the computation."

Name _____

Study the example and complete each move.

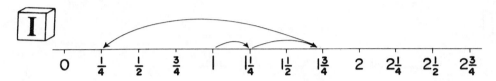

| 0 | ¼ | ½ | ¾ | 1 | 1¼ | 1½ | 1¾ | 2 | 2¼ | 2½ | 2¾ |

Start Go   Go   Go   Stop

$\boxed{1} \rightarrow \left(\frac{1}{4}\right) \rightarrow \left(\frac{1}{2}\right) \leftarrow \left(1\frac{1}{2}\right) = \left(1\frac{1}{4}\right)$

1. $\boxed{\frac{1}{4}} \rightarrow \left(\frac{3}{4}\right) \rightarrow \left(\frac{1}{4}\right) \rightarrow \left(\frac{1}{2}\right) = \bigcirc$

2. $\boxed{\frac{3}{4}} \rightarrow \left(\frac{1}{2}\right) \leftarrow \left(\frac{3}{4}\right) \rightarrow \left(\frac{1}{2}\right) = \bigcirc$

3. $\boxed{1\frac{1}{2}} \rightarrow \left(\frac{3}{4}\right) \leftarrow \left(1\right) \rightarrow \left(\frac{1}{4}\right) = \bigcirc$

4. $\boxed{1\frac{3}{4}} \rightarrow \left(\frac{1}{4}\right) \leftarrow \left(\frac{1}{2}\right) \leftarrow \left(\frac{3}{4}\right) = \bigcirc$

5. $\boxed{2} \leftarrow \left(\frac{1}{2}\right) \leftarrow \left(\frac{3}{4}\right) \rightarrow \bigcirc = \left(1\right)$

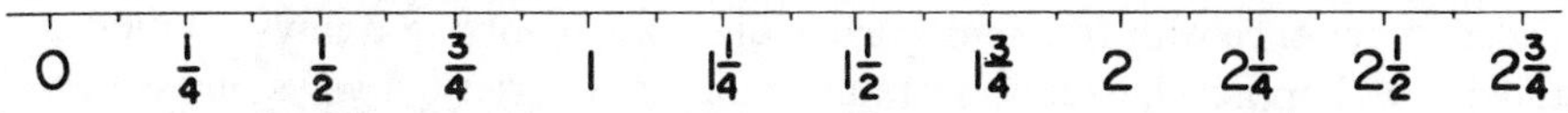

| 0 | ¼ | ½ | ¾ | 1 | 1¼ | 1½ | 1¾ | 2 | 2¼ | 2½ | 2¾ |

6. $\boxed{1} \leftarrow \left(\frac{1}{8}\right) \rightarrow \left(\frac{3}{8}\right) \rightarrow \left(\frac{5}{8}\right) = \bigcirc$

7. $\boxed{} \rightarrow \left(\frac{7}{8}\right) \rightarrow \left(\frac{1}{2}\right) \rightarrow \left(\frac{1}{4}\right) = \left(2\right)$

8. $\boxed{\frac{3}{4}} \rightarrow \bigcirc \leftarrow \left(\frac{3}{8}\right) \leftarrow \left(\frac{3}{8}\right) = \left(1\frac{3}{4}\right)$

9. $\boxed{} \rightarrow \left(1\frac{1}{2}\right) \leftarrow \left(\frac{3}{4}\right) \rightarrow \left(\frac{1}{2}\right) = \left(1\frac{1}{2}\right)$

10. $\boxed{\frac{7}{8}} \leftarrow \left(\frac{1}{2}\right) \rightarrow \left(\frac{5}{8}\right) \rightarrow \bigcirc = \left(1\frac{1}{2}\right)$

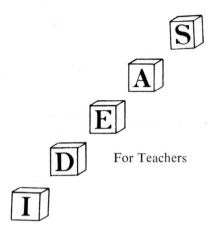

For Teachers

Objective: Number line experience with fractions

Grade level: 6, 7, or 8

Directions for teachers:

1. Remove the worksheet and reproduce a copy for each student.
2. Have students study the example and discuss it.
   $$\left(\begin{array}{c}\text{START GO} \\ \boxed{m} \rightarrow \boxed{n}\end{array} \text{ is read: "Start at } m \text{ and move } n \text{ to the right."}\right)$$
3. Have students go ahead on their own.
4. Anticipate that exercises 7, 8, and 9 will cause considerable frustration. Challenge your students to figure out solutions they can "defend."
5. Discuss the various methods of solution devised by your students.

Comments: These exercises provide experiences with equivalence of fractions (i.e., $\frac{1}{2} = \frac{4}{8}$), inverses (i.e., $\rightarrow\textcircled{$\frac{1}{2}$}\leftarrow\textcircled{$\frac{1}{2}$} = 0$) as well as the much needed experience with fractions greater than one involving a physical model. Students will approach exercises of this nature in a variety of ways unless they are told how to think or do. The student who needs counting experiences on the number line will have that experience. The student who has good insight into the patterns of mathematics will devise far more efficient procedures.

This experience may be expanded in many ways: Other special subsets of fractions may be chosen, such as thirds and sixths; decimal-fraction number lines could be used; and equations with two variables ($\textcircled{$\frac{1}{4}$}\rightarrow \bigcirc \leftarrow \bigcirc \rightarrow\textcircled{$\frac{1}{2}$} = \textcircled{1}$) would be challenging.

Answers

1. $1\frac{3}{4}$	3. $1\frac{1}{2}$	5. $\frac{1}{4}$	7. $\frac{3}{8}$	9. $\frac{1}{4}$
2. 1	4. $\frac{3}{4}$	6. $1\frac{7}{8}$	8. $1\frac{3}{4}$	10. $\frac{1}{2}$

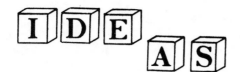

MERRY MEASURING

Cut a piece of string equal to your height.

Fold it in half and try it on yourself.

What can you find that is $\frac{1}{2}$ your height?

Fold it in thirds. What can you find that's $\frac{1}{3}$ your height? $\frac{1}{4}$? $\frac{1}{5}$? What else?

RECORD HERE →

$\frac{1}{2}$ my height	$\frac{1}{3}$ my height	$\frac{1}{4}$ my height	$\frac{1}{5}$ my height	What else?

For Teachers

Objective: Experience with body measures and body ratios.

Levels: 3, 4, or 5

Directions for teachers:

1. Remove the activity sheet MERRY MEASURING and reproduce one copy for each student.

2. Discuss it to make sure the directions are understood.

3. Give each child a piece of string. Make sure the string doesn't have any stretch.

Follow-up:

Additional problems using the piece of string can be suggested.

(1) How many of your widest smile make your height? Guess first, then use your string.

(2) Did your mother ever wrap a sock around your fist to see if it was your size? Why would she do a thing like that? Use your string to see.

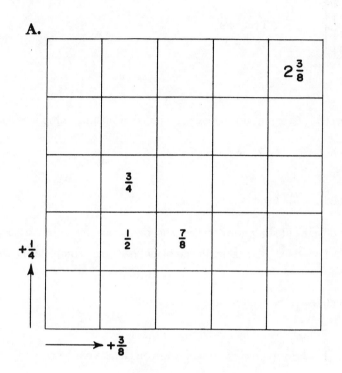

Follow the arrows.

A.

				$2\frac{3}{8}$
	$\frac{3}{4}$			
$+\frac{1}{4}$ ↑		$\frac{1}{2}$	$\frac{7}{8}$		

→ $+\frac{3}{8}$

B.

	$\frac{2}{5}$			
$\times\frac{1}{5}$ ↑		2	1	

→ $\times\frac{1}{2}$

For Teachers

Objective: Computational practice with addition and multiplication of fractions.

Grade level: 6, 7, or 8

Directions for teachers:

1. Tear out student worksheet and reproduce one copy for each student.
2. Draw a similar table on the blackboard and complete the table in class.

Directions to be read to students:

1. Follow the arrows to complete the table.
2. Look for patterns that will help you check your work.

Comments: Tables like these are easily made up and give computational practice that is fun. You may also wish to explain the meaning for other arrows such as ↗, ↘, or ↙, or make combinations of arrows and interpret these arrows and their combinations in terms of fraction operations.

Put an X on each solution of the inequality, 1.0 + ☐ > 2.5

Connect the X's and find the path from START to FINISH.

	START			
1.4	1.6	6.5	2.1	1.7
0.8	0.0	0.6	1.2	4.0
2.7	2.2	1.8	2.3	2.9
3.0	0.2	0.9	1.1	0.4
2.0	4.6	3.5	1.0	1.5
		FINISH		

For Teachers

Objective: To practice adding and subtracting decimals by finding solutions to an inequality

Levels: 6, 7, 8

Directions for teachers:

1. Give each student a copy of the worksheet.

2. Let the students read the directions.

3. If students have trouble finding solutions to the inequality, ask them to add 1.0
to each number in the array and to compare their answers with 2.5.

IDEAS

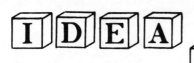

Name ______________________

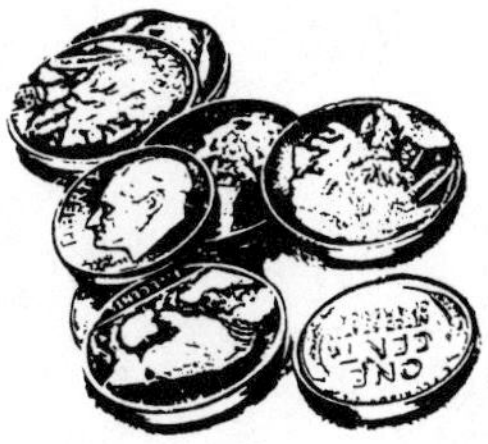

How many different ways can you have 25¢?

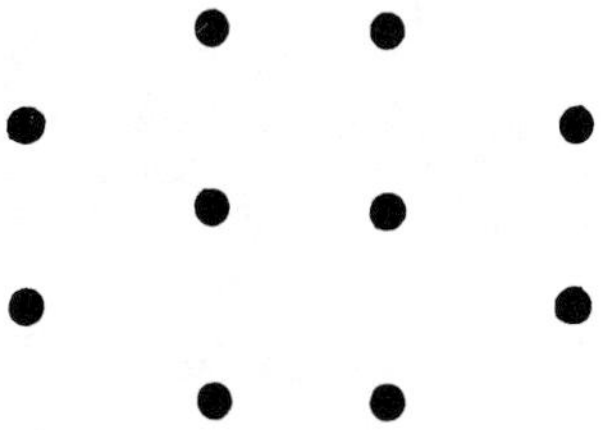

You have 10 points. How many line segments are needed to connect each pair?

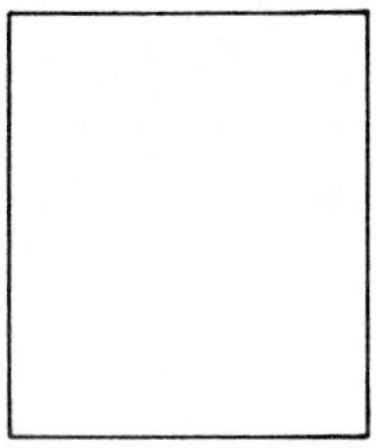

How many ways can you fold a piece of paper in half?

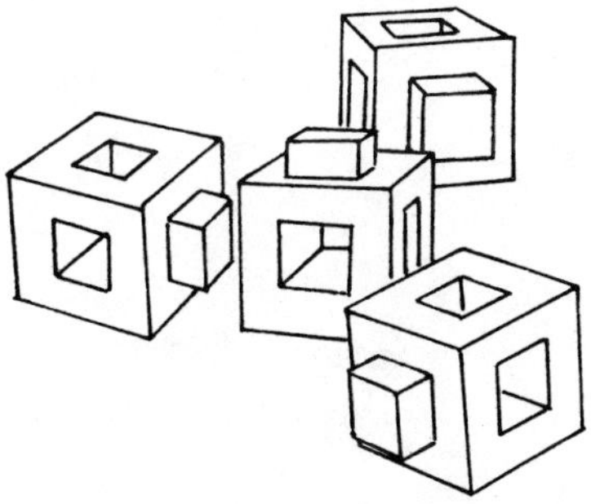

How many shapes can you make with 4 cubes?

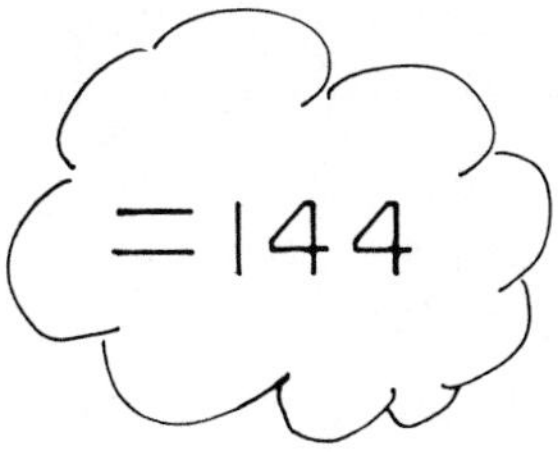

Write 10 computational exercises that have the answer 144. Use +, −, x, and ÷, at least once in each exercise.

Levels 3 or 4

Objective: Experiences in problem solving

Directions for teachers: (all levels)

1. Remove the activity sheet and make a copy of it.
2. Cut the problems apart and paste each one on a 5-by-7 card.
3. Suggestions for use:
   a. Post one as the "Problem of the Week." Post student solutions with next week's problem.
   b. Give one to each team of students. Have teams report their progress or solution.
   c. Give one to an individual as a special challenge or a special project.

Comments: Be receptive to partial solutions and incomplete reasoning patterns. Encourage students to test their ideas. Open-ended problems such as these often suggest other problems to the perceptive student. Encourage your students to create problems for your file.

—25¢: 8 ways
—10 points: 45 line segments
—Endless number of ways. Any fold through point P.

—4 cubes: 8 shapes
—Examples: $100 \div 2 \times 3 + 10 - 16 = 144; 75 \div 3 \times 8 - 70 + 14 = 144$

Mint chip

Vanilla

Chocolate

How many ways can you arrange the three scoops of ice cream? Draw them.

How many ways can you arrange one scoop? Two scoops?

Complete this table

Scoops	Ways
1	
2	
3	
4	
5	

Can you find a way to figure the pattern? Hint: It helps to multiply.

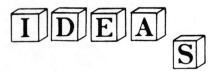

Levels: 4, 5, 6

Objective: To investigate permutations—an activity in logical thinking

Directions for teachers:

Duplicate a worksheet for each child. The worksheet is self-explanatory, but make sure the children understand the directions.

The solution:

Scoops	Ways
1	1
2	2
3	6
4	24
5	120

Mathematically, the rule has to do with the concept of "factorial." For example, 3! is read "three factorial" and is the product of the integers from 1 to 3.

$$3! = 3 \times 2 \times 1 = 6$$
$$4! = 4 \times 3 \times 2 \times 1 = 24$$

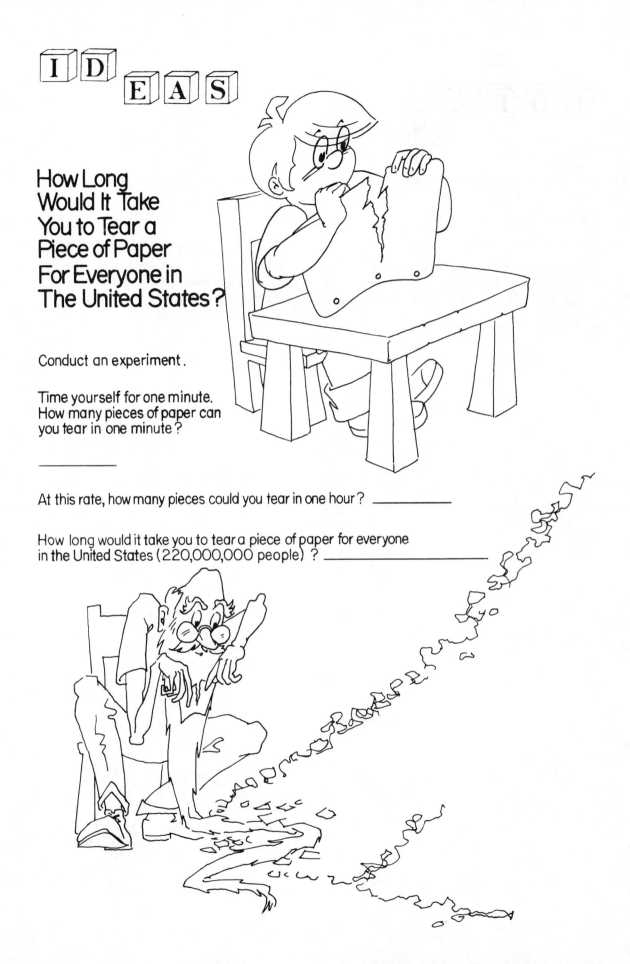

IDEAS

How Long Would It Take You to Tear a Piece of Paper For Everyone in The United States?

Conduct an experiment.

Time yourself for one minute. How many pieces of paper can you tear in one minute?

_____

At this rate, how many pieces could you tear in one hour? _____

How long would it take you to tear a piece of paper for everyone in the United States (220,000,000 people)? _____

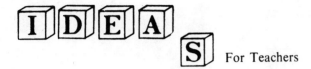 For Teachers

Objective: Experience in problem solving using large numbers

Levels: 4, 5, 6, 7, 8

Directions for teachers:

1. Remove the activity sheet and reproduce a copy for each student.
2. To solve the problem, have students conduct an experiment by following the directions.

Comments:

If calculators are available, students should be encouraged to use them. The main focus of this activity is on problem solving rather than computation.

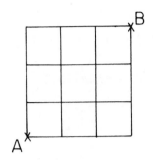

This is a map for a housing development. Your task is to plan a bus route from A to B. How many possible routes are there? Which is the shortest?

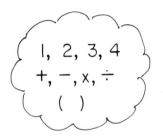

1, 2, 3, 4
+, −, x, ÷
( )

Use only these symbols to name the numbers from 1 to 25. The 1, 2, 3, and 4 must be used once and only once in naming a number. The other symbols may be used as often as you wish.

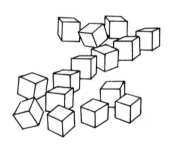

How many different prisms can you build using <u>no more</u> than 15 cubes?

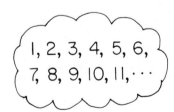

1, 2, 3, 4, 5, 6, 7, 8, 9, 10, 11, ···

Using only the counting numbers, how many solutions are there to these equations?

$$x + y = 10 \qquad x + y = 25$$
$$x + y = 100$$

Place 3 nails so they are all the same distance apart. Now try the same task with 4 nails.

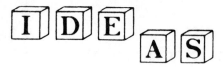

Levels 5 or 6

Objective: Experiences in problem solving

Directions for teachers: (all levels)

1. Remove the activity sheet and make a copy of it.
2. Cut the problems apart and paste each one on a 5-by-7 card.
3. Suggestions for use:
   a. Post one as the "Problem of the Week." Post student solutions with next week's problem.
   b. Give one to each team of students. Have teams report their progress or solution.
   c. Give one to an individual as a special challenge or a special project.

Comments: Be receptive to partial solutions and incomplete reasoning patterns. Encourage students to test their ideas. Open-ended problems such as these often suggest other problems to the perceptive student. Encourage your students to create problems for your file.

Answer Key

—45 routes with no backtracking. All routes: 6 blocks.
—$(2 + 3) \div (4 + 1) = 1$; $(4 \times 2) \div (3 + 1) = 2$; $(4 \div 2 - 1) \times 3 = 3$; etc.

15 by 1	7 by 2	5 by 3	2 by 2 by 2
14 by 1	6 by 2	4 by 3	2 by 2 by 3
13 by 1	5 by 2	3 by 3	
12 by 1	4 by 2		
11 by 1	3 by 2		
.	2 by 2		

.

.

1 by 1
26 prisms
—9, 24, 99
—The fourth nail must be above these three thereby forming the points of a regular tetrahedron.

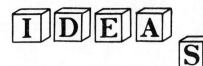

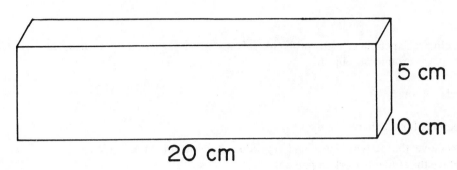

5 cm

10 cm

20 cm

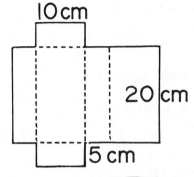

10 cm

20 cm

5 cm

Wrap the Christmas present.

1. How much wrapping paper is needed to cover the box? _____

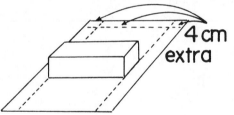

4 cm extra

2. Allow 4 centimeters for an overlap going around the box and 4 centimeters extra on each end. How large a piece of paper do you need? _____

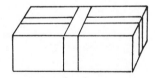

3. Tie it with a ribbon. How long a ribbon do you need to go around both ways?

_____

4. Allow 50 centimeters for a bow. How much ribbon do you need? _____

Research Problem: Work out a rule for the amount of paper needed to wrap any box (allow a 4 cm. overlap and ¾h. on each side to cover the ends).

h

w

l

 For Teachers

Objective: Experience with problem solving involving area and distance around an object.

Levels: 4 or 5

Directions for teachers:

1. Remove the activity sheet and reproduce a copy for each student.
2. Have the students work in pairs.
3. Provide ¼-inch or ½-inch graph paper for those that need to make a model or do a layout of the model.
4. Post solutions to the "Research Problem."

Comments: Problem solving is the primary purpose of mathematics. As our curricula move toward an empahsis on applied mathematics, we will experience an increased focus on problem-solving skills. Children need a wide variety of problem-solving experiences to build their skills.

Key: 1. 700 sq. cm. 2. 28 cm. by 34 cm. or 952 sq. cm. 3. 80 cm. 4. 130 cm. Research Problem: $(2w + 2h + 4)$ by $(l + 3/2h)$

Partners _____

_____

1. How big is the package?
   • Its length is twice its width.
   • Its height is half its width.
   • Its length is 25 centimeters.

   length _____
   width _____
   height _____

2. How much does it weigh?
   • One fourth of the package weighs
     200 grams less than one half of
     the package.

   weight _____

3. How much did it cost?
   • If you buy two of them, you
     would get $4.50 change from
     a $20 bill.

   cost _____

4. Who is it for? Who is it from?
   Children: Todd, Steve, Mark, Jill
   • It is not from a boy to the girl.
   • It is not from a boy to a boy.
   • It is not for Todd or Mark.

   For _____

   From _____

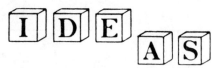 For Teachers

Objective: Experience in problem solving where elementary logic is used.

Levels: 5 or 6

Directions for teachers:

1. Remove the activity sheet and reproduce a copy for each student.
2. Have students work in pairs.
3. If necessary suggest using the "guess and test" approach or making a table. Do not impose any formal procedures on the student.
4. You may wish to have successful students tell how they solved the problem.

Comments: Problem solving is the primary purpose of mathematics. As our curricula move toward an emphasis on applied mathematics, we will experience an increased focus on problem-solving skills. Children need a wide variety of experiences to build their problem-solving skills.

Key: 1. 25 × 12½ × 6¼. 2. 800 grams. 3. $7.75. 4. Steve, Jill.

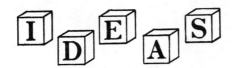

Name _____

Name _____

1. Choose a page from the phone book.

2. Write down the sum of the last two digits of 50 telephone numbers. 266-72⑰ → 8

Trials

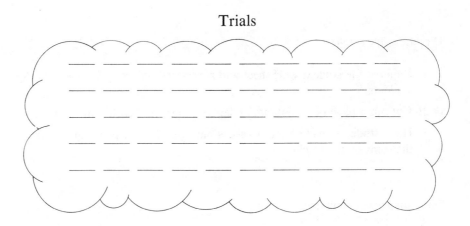

Sum	Tally	Total
0		
1		
2		
3		
4		
5		
6		
7		
8		
9		

Sum	Tally	Total
10		
11		
12		
13		
14		
15		
16		
17		
18		
19		

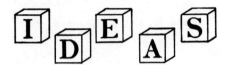

For Teachers

Objective: Experience in reading data.

Grade level: 4, 5, 6, 7, or 8

Directions for teachers:

1. Remove the student worksheet and reproduce one copy for each pair of students.
2. Cut out a page of an outdated telephone book for each pair of students.
3. Have students read phone numbers and identify last two digits, and put the sums on the blackboard.

Directions for students:

1. You are to work in pairs.
2. Each pair should mark 50 numbers in a row on your page of the phone book.
3. List the sum of the last two digits on the blanks on your worksheet.
4. Make a tally for each sum in the table and find the totals.

Comments: Many students may need some instruction when it comes to tallying their data. You may want to stop the activity when the students reach this point and explain this technique. Depending upon the level of your class, you may want to discuss the most likely sum, the average, the median, or the mode of the resulting frequency. You may also want to compile a class frequency from the individual data. Most students will be anxious to explain why no tally marks appear after "19."

Name _____

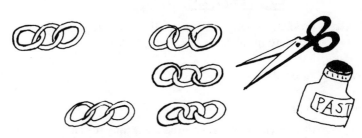

1. How many links do you need to cut and paste together to make one 15 link "chain" out of these 5 pieces? _____

2. How can you give one of these four presents to each of 4 girls so that one present is left in the box? _____

3. A boy bought one present for each of his sisters. He also bought one present for each of his brothers. He bought as many presents for brothers as he did for sisters. His sister also bought one present for each. She bought only half as many presents for sisters as she did for brothers. How many brothers does he have?

_____

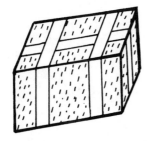

4. The present belongs to Bill, Ed, or Jon. Whose present is it? _____

These statements are true:
   1. The present belongs to Bill or Ed.
   2. The present belongs to Ed or Jon.
   3. The present does not belong to Jon.

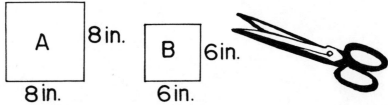

Cut squares A and B each into 2 pieces. Fit the 4 pieces together to form one 10 inch by 10 inch square.

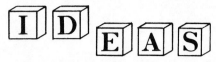 For Teachers

Objective: Problem-solving experiences that encourage the students to employ a variety of methods of solution.

Levels: 6, 7, or 8

Directions for teachers:

1. Remove the activity sheet and reproduce a copy for each student.
2. Present these problems as a challenge. Seek a variety of solutions rather than "the" answers.
3. Be as receptive to good thinking and innovative approaches as correct solutions.
4. Post detailed written solutions or have successful students explain their solutions to others.

Comments: Neither encourage nor discourage students working collectively on these problems. They present a level of challenge that is best approached through the student's chosen learning style.

Key:

1. Three links need to be cut.

2. Give a present to each of 3 girls, then give the box with the remaining present to the fourth girl.
3. 3 brothers (also 3 sisters)
4. Ed. (Several reasoning patterns are possible.)

5. (Other solutions are possible.)

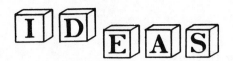

Team _____

_____

_____

_____

1. Find an example of a parallelogram in the classroom.

2. Draw a polygon whose area is the same as its perimeter.

3. Fold a piece of paper so that the sum of the angles formed is 180°.

4. Draw a polygon whose diagonals intersect at right angles.

5. Draw a rectangle and a triangle that have the same area.

6. Find 4 different numbers whose sum is greater than their product.

7. Find the smallest number that is evenly divisible by 2, 3, 4, and 5.

8. Find a number for which the sum of all the factors is twice the number.

9. Describe all the points that are 1 inch from point A.

•A

 For Teachers

Objective: A team experience in problem solving

Levels: 6, 7, or 8.

Directions for teachers:

1. Remove the activity sheet and reproduce a copy for each team.
2. Form teams of four students each.
3. Introduce the team to the rules for the "Scavenger Hunt."
   a) It is a team task.
   b) The team that turns in the most correct tasks in the time allotted is the winner.
   c) In case of a tie, the first team to turn in their tasks is the winner.
4. Discuss solutions.
5. Discuss various strategies used by the teams.

Comments: Be especially observant of the strategies used as the teams participate in the "Scavenger Hunt."

Key: 1) Any rectangle qualifies. 2) Many solutions are possible. The simplest are a 4-by-4 square or a 3-by-6 rectangle. 3) Typical solutions:

(each represents a different interpretation of the task). 4) Any quadrilateral with four equal sides—a square is the most common. 5) Many solutions. 6) Any 3 numbers combined with zero, combinations of fractions, or some combinations of positive and negative numbers. 7) 60. 8) 6, 28, (any "perfect" number). 9) A sphere with radius of 1 inch.

Name _____

Ring those words that name
the corners of a triangle :

ROD   RED

RAN   DUG

CAN   YOU

NOD   DOG

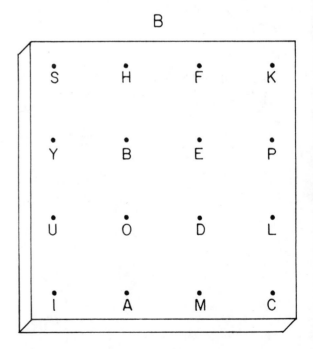

List other words that name corners of a triangle.

___ ___ ___ ___ ___

Ring those words that name
the corners of a square.

SICK   FOAM

HUMP   FLAY

SULK   HACK

MODE   DAUB

List words that name corners of a rectangle.

_____   _____

Objective: Experiences with basic geometric shapes and the standard technique for naming the shapes.

Levels: 3, 4, or 5

Directions for teachers:

1. Remove the activity sheet and reproduce a copy for each student.

2. Encourage students to visualize the triangles without actually drawing them. (Expect that some students will have to draw the triangles before they can see them.)

3. There are many 3-letter words on geoboard A that name the corners of a triangle. Some students will list more than the number of blanks shown.

4. The letters naming the corners of a square must be in clockwise or counterclockwise order. (MOLE is a word but the square would not be properly named with the letters in that order.)

5. The student who knows that all squares are rectangles may fill the last blanks with names for squares.

Comments: You may wish to hand out 4-by-4 arrays of dots challenging the students to place letters on them to form squares, rectangles, or other polygons.

Key:

A: ROD, CAN, NOD, DUG, DOG

B: SICK, HUMP, FLAY, DAUB

Name _____

Find a polygon for each condition:

1. A right triangle that chops wood.
   ____ ____ ____

2. A triangle that names man's best
   friend. ____ ____ ____

3. A square that names a man.
   J ____ ____ ____

4. A parallelogram that grows on
   ears. ____ ____ ____ ____

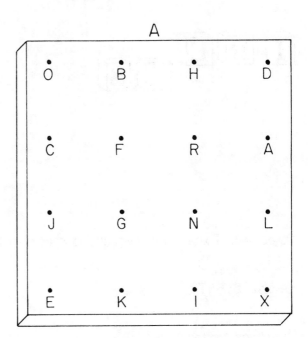

5. A parallelogram that is used to gather leaves. ____ ____ ____ ____

6. A trapezoid that usually has frosting. ____ ____ ____ ____

Find a polygon for each condition:

1. An isosceles triangle that flies
   at night. _____

2. A right triangle to wipe your
   shoes on. _____

3. An isosceles right triangle that
   is done on a chair.

   _____

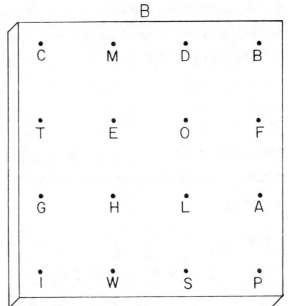

4. An obtuse triangle that is placed in a horse's mouth. _____

5. A rectangle that holds a sail. _____

6. A concave polygon that means : to brag. _____

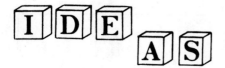

For Teachers

Objectives: Experience in visualizing special polygons

Levels : 5,6,7

Directions for teachers :

1. Remove the activity sheet and reproduce a copy for each student .
2. Encourage students to visualize the polygons without actually drawing them .
3. Annouce that a key will be posted at a specific time .
4. Polygons are lettered clockwise or counterclockwise. (Normally B A N D is not considered a polygon .)

Comments : Hand out 4-by-4 dot arrays that do not have letters. Challenge your language oriented students to make up similar sets of questions.

Key :
  A : 1. AXE 2. DOG 3. JOHN 4. CORN
          5. RAKE 6. CAKE
  B : 1. BAT 2. MAT 3. SIT 4. BIT
          5. MAST 6. BOAST

IDEAS

On geoboard A:

Draw all the polygons.
Use all the dots on the
geoboard.

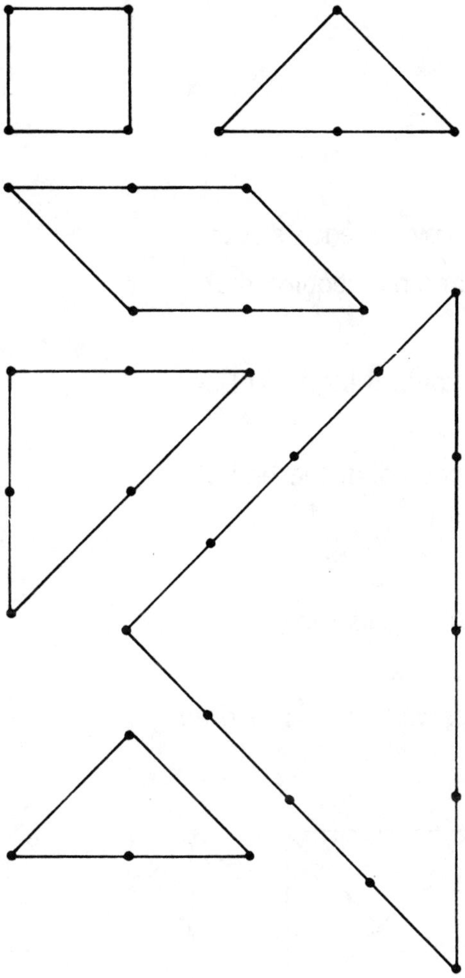

Geoboard A

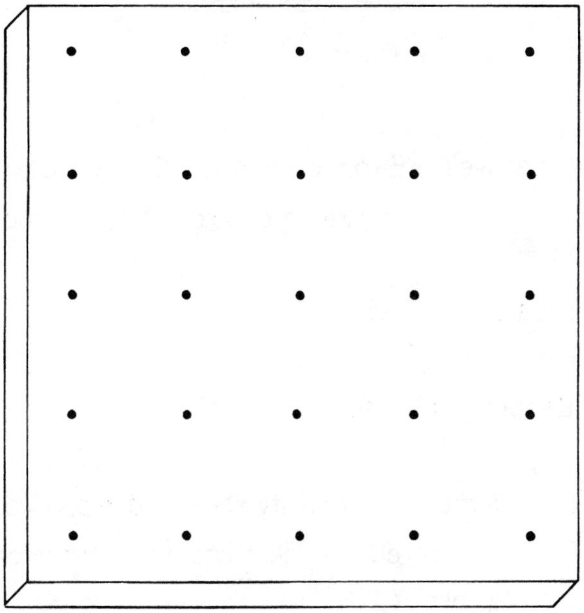

Geoboard B

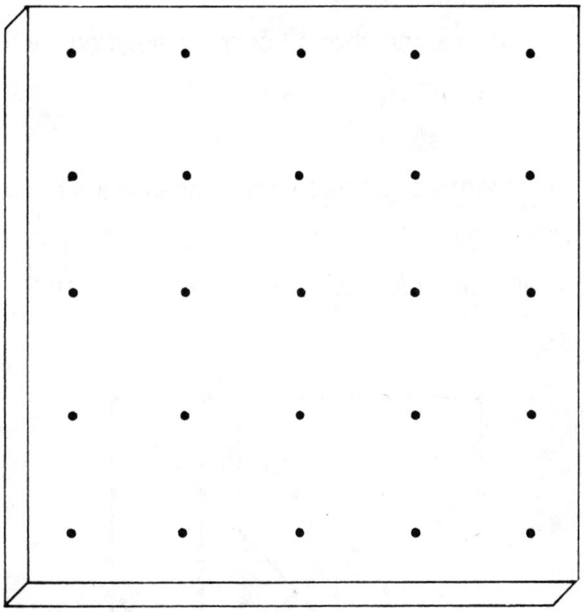

On geoboard B:

a) Draw one of the above polygons twice to form a right triangle.

b) Draw one of the polygons twice to form a square.

c) Draw one of the polygons twice to form a parallelogram.

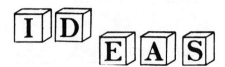 For Teachers

Objectives : Experience in visualizing and drawing composite polygons that
requires a concept of congruence.

Levels : 6, 7, or 8

Directions for teachers :

1.   Remove the activity sheet and reproduce a copy for each student.
2.   Be sure each student has a straight edge but do not require that
he use it.
3.   Present this activity as a challenge. Don't expect a high level of
success.
4.   Announce that all correct solutions will be posted at the end of
one week.

Comments : These experiences are similar to but considerably more
challenging than their counterpart using tangrams. You may wish to
extend this idea into the study of different shapes with the same area.

Key :

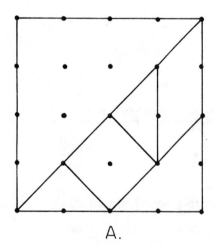
A.

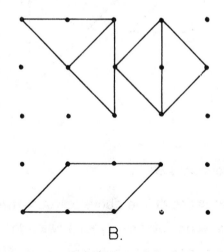
B.

IDEAS

Name ____________________

Use the points as corners:

- draw a square
- draw a rectangle
- draw a triangle with equal sides.

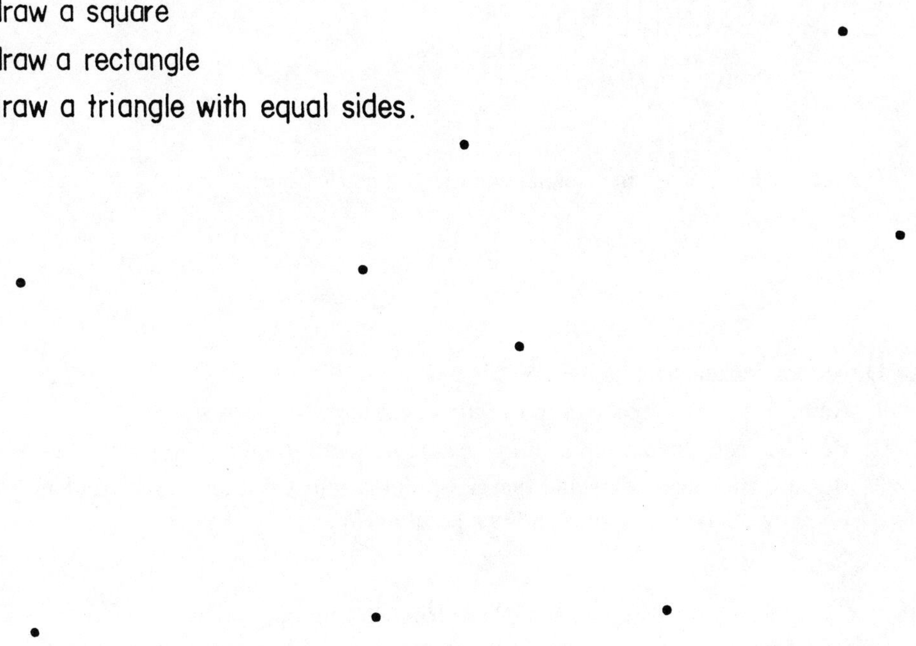

Draw 3-point line segments:

- that are parallel
- that form a "T".
- that form a triangle.

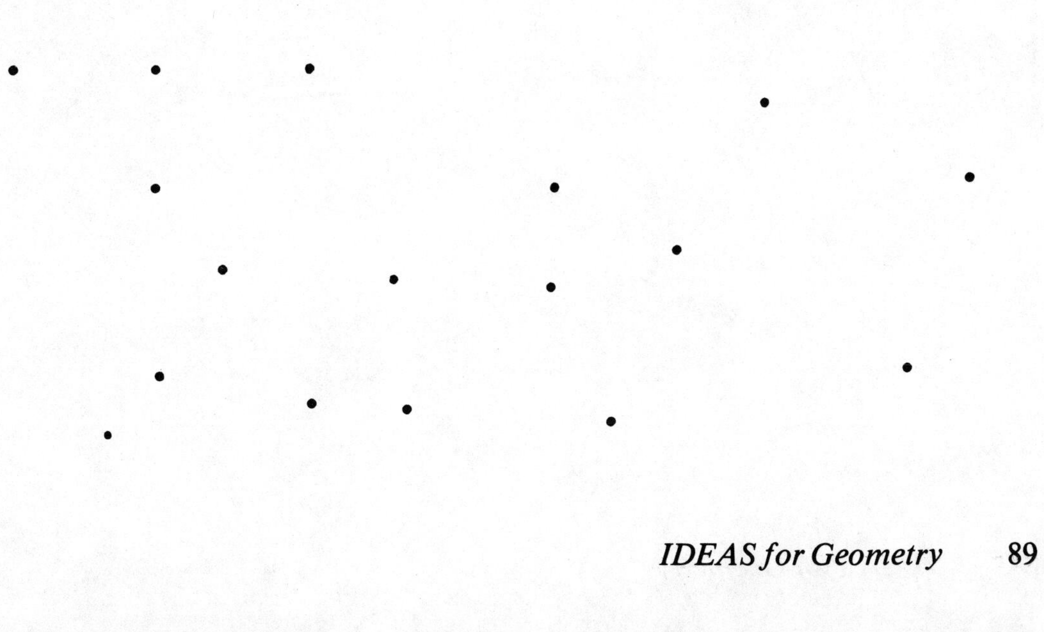

For Teachers

Objective: Experience in visualizing basic geometric objects

Levels: 3 or 4

Directions for teachers:
1. Remove the activity sheet and reproduce a copy for each student.
2. Be sure each student has a straightedge but do not require that he use it.
3. Be sure students understand that a 3-point segment is a line segment identified by the two endpoints and one point between.

Comments: Visualizing a rectangle or triangle with only the vertices (corner points) shown is a far more challenging experience than assigning a name to a picture. Some highly number-oriented students may experience frustration, while some of your nonnumerically oriented students may exhibit exceptional insight in spatial relations.

Key:

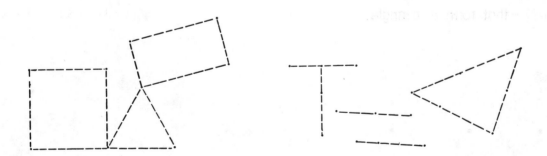

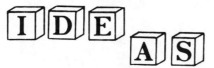

Use only the points below as corners:

Draw a square whose area is 16 square centimeters.

Draw a rectangle whose area is 33 square centimeters.

Draw the largest square you can. What is it's area ? _____

Use the points below to draw 3 squares.

Name them . _____ , _____ , _____ .

A .

.B

C
.

D

E .

.H

F .

. G

I
.

. K

J .

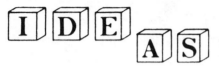

For Teachers

Objective: Experience in visualizing specific geometric objects

Levels: 4, 5, or 6

Directions for teachers:

1. Remove the activity sheet and reproduce a copy for each student.
2. Be sure each student has a centimeter rule.
3. As you observe students working, be sure they understand that only the given points can be used as "corners" and that not all points need to be used.

Comments: Since it is *not* the purpose of this activity to teach the concepts, care should be taken that this sheet is used only after the student has studied the area of squares and rectangles using metric measures.

Key:

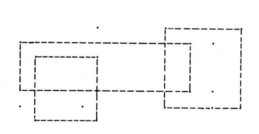

ABGF
DCKJ
GEHI

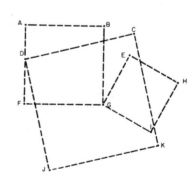

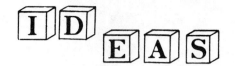

Use the points below as corners

      Draw two congruent rectangles.

      Draw two congruent <u>obtuse</u> triangles.

Use the points below as corners

      Draw two rectangles with equal areas.

      Draw two congruent <u>acute</u> triangles.

For Teachers

Objective: Experience in visualizing geometric objects with specific characteristics

Levels: 6, 7, or 8

Directions for teachers:

1. Remove the activity sheet and reproduce a copy for each student.
2. Be sure that each student has a straightedge but do not require that he use it.
3. Have students work independently.

Comments: The activities can be made considerably more challenging by having your students try to find the congruent triangles *before* they draw the rectangles or by adding one or two extraneous points. The latter must be done very carefully so that other pairs of solutions are not introduced.

Key:

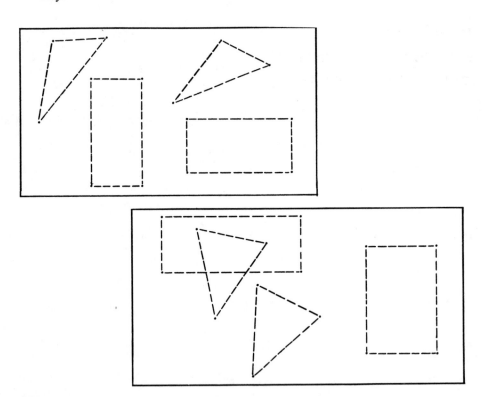

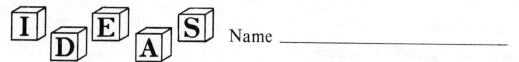
Cut out these pieces and make two squares.

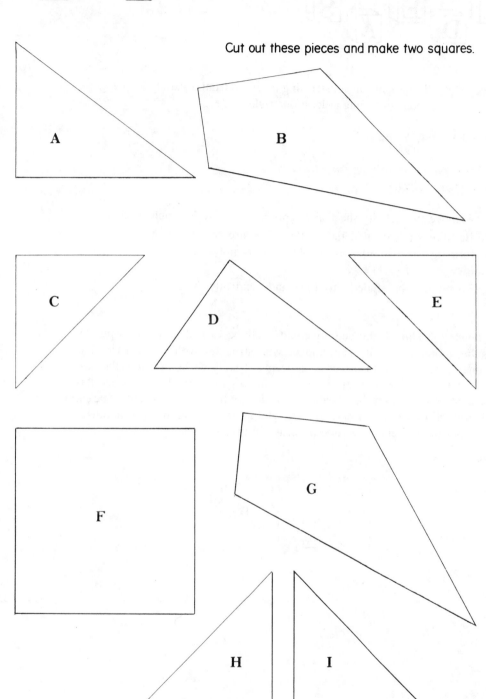

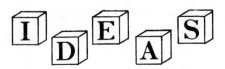

Objective: Experience in constructing a square from a variety of polygons, each containing at least one right angle

Grade level: 4, 5, or 6

Directions for teachers:

1. Remove the activity sheet and reproduce a copy for each student.
2. Be sure students understand that they are to use all nine pieces in forming the two squares. (Polygon *F* is not one of the two squares they are to form!)
3. Encourage the students to work independently.

Comments: Though this activity could easily be viewed simply as a puzzle, it is far more than that. The student who struggles with this activity has personal experience with the basic concepts of congruence and tessellations. The fact that the solution produces two squares that are the same size has important though subtle implications for the sophisticated concept of area. A skillful discussion leader may be successful in drawing out many generalizations from the students if he doesn't insist on precise language.

Solution

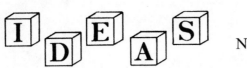
Name _____

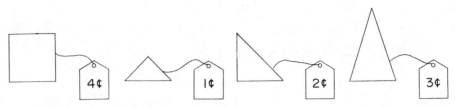

How much are these?

A. 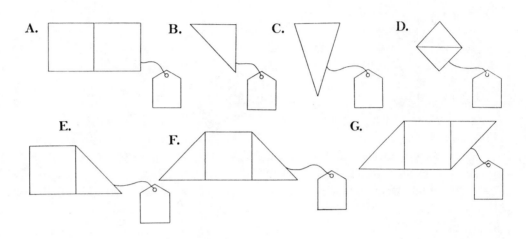   B.   C.   D.

E.   F.   G.

H.   I.   J.

K.   L.

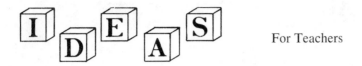 For Teachers

Objective: Experience with identification of congruent polygons

Grade level: 1, 2, 3, 4, 5, 6, 7, or 8

Directions for teachers:

Remove the student worksheet and reproduce a copy for each student.

For grades 1 and 2:
1. Discuss the square and triangles shown at the top of the page. Be sure to point out their cost.
2. Discuss the cost of the first four or five examples and fill in the tags.
3. Let the students do the rest of the examples on their own.

For grades 3, 4, 5, 6, 7, and 8
1. Have the students study the polygons at the top of the page and complete the price tags for the other polygons.
2. After they have completed the worksheet, discuss the relation between those polygons that are the same price. The area concept should come out in the discussion.

Comments: Fundamental to the development of many area concepts is the idea of conservation. In this case the area of a polygon does not change as we move it around or place it with other polygons. The use of price provides a different focus on area and forces the student to consider area in a different way. A variety of approaches to the development of a concept broadens the concept for some students and develops understanding for students that didn't see the idea before.

Answers

A. 8   B. 2   C. 3   D. 2   E. 6   F. 8   G. 8   H. 8   I. 4   J. 8   K. 16

Anticipate both 16 and 20 as answers for L.

Name _____

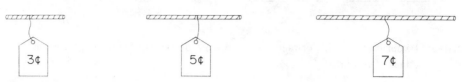

How much are these?

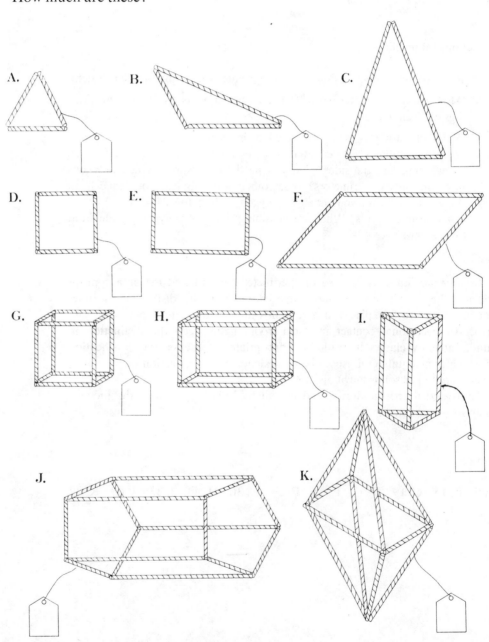

 For Teachers

Objective: Experiences with perimeter of polygons and identification of the edges of a solid

Grade level: 3, 4, 5, 6, 7, or 8

Directions for teachers:

1. Remove the student worksheet and reproduce one copy for each student.

2. After handing out the worksheet, ask the students to fill out the price tags on each figure.

3. When the students have completed their answers, discuss the different ways the students arrived at the answers: How did you know which straws make up the sides? Did you need to measure? Which polygons have the largest perimeters? What other polygons can you make from these straws? Can you make a polygon selling for 21 cents? For 13 cents? For 28 cents? What are possible prices for polygons made from these straws?

Comments: Students confuse the perimeter concept and the area concept because they don't have enough experience where the distinction is functional. There are few places in a student's life where he uses perimeter and area. An occasional contact in a classroom helps keep the distinction in mind. In many classes it would be appropriate to discuss the classification of triangles as equilateral, isosceles, or scalene. An investigation of pyramids, prisms, and other solids might also result.

Third and fourth graders would benefit by building some of the models, using straws and tape.

Answers

A. 9   B. 15   C. 19   D. 12   E. 16   F. 24   G. 36   H. 44   I. 33   J. 65   K. 72

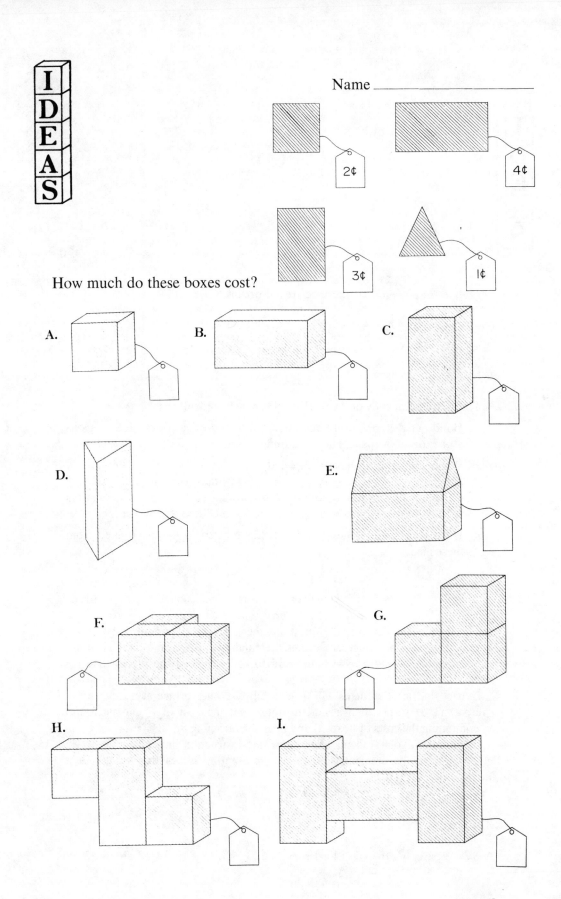

How much do these boxes cost?

For Teachers

Objective: Experience with surface area of prisms

Grade level: 5, 6, 7, or 8

Directions for teachers:

1. Reproduce a copy of the worksheet for each student.

2. Hand out the copies and call attention to the prices of the four regions at the top of the page. Have the students figure out the price of each box.

3. When the students have completed the sheet, discuss their methods of arriving at the answers. Pay particular attention to F and G, which are different views of the same box. There are at least two possible answers for I, depending on whether the student "cuts" the 2¢ pieces.

4. Have your students design other boxes that could be constructed from these pieces.

Comments: Since the student lives in a three-dimensional world, he needs experiences in three-dimensional geometry. These activities provide experience in visualization. To answer the questions, he is forced to visualize the "back side" and to see the congruent pieces that make up the box. Boxes H and I are real challenges both to visualize and to keep track of the pieces that are used.

Expect that most students will figure each box one surface at a time; however, don't be surprised if a student notes that the cost of E can be found by combining the costs of boxes B and D and subtracting 8¢.

Even though your students won't have had instruction in perspective drawing, some will be amazingly adept at representing boxes they've designed themselves.

Answers

A. 12　B. 20　C. 16　D. 14　E. 26　F. 28　G. 28　H. 36　I. 56 or 52

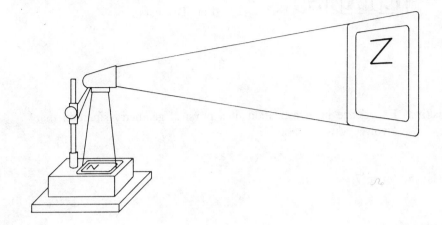

Objects   Length of image of

a   a. _____
_____

b   b. _____
_____

c   c. _____
_____

d

Perimeter of image of d. _____ ;

e. _____ ;

Area of image of d. _____ ;

e. _____ .

e

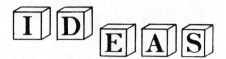

For Teachers

Objective: Experience in investigation of a physical geometry situation that involves constant ratio.

Levels: 5, 6, or 7

Directions for teachers:

1. Remove the activity sheet. Make a copy for each student and a transparency for your use.
2. Make sure that each student has a ruler.
3. Before class, set your projector so that the image of line segment *a* is 5 inches long.
4. Have each student measure segment *a* on his paper. Then have a volunteer measure its image on the screen (or chalkboard).
5. Have each student measure *b* on his paper. Then have each student guess how long the image of segment *b* is.
6. Repeat this process (step 5) with segment *c* and the perimeters and areas of the squares *d* and *e*.

Comments: The activity can be repeated with any desired setting of the projector. If the image of segment *a* is 10 times the length of *a*, then the image of segment *b* will be 10 times the length of *b*, and so on. The fact that segment *b* is twice the length of segment *a*, and that segment *c* is three times the length of segment *a* provides a second pattern for the student to investigate.

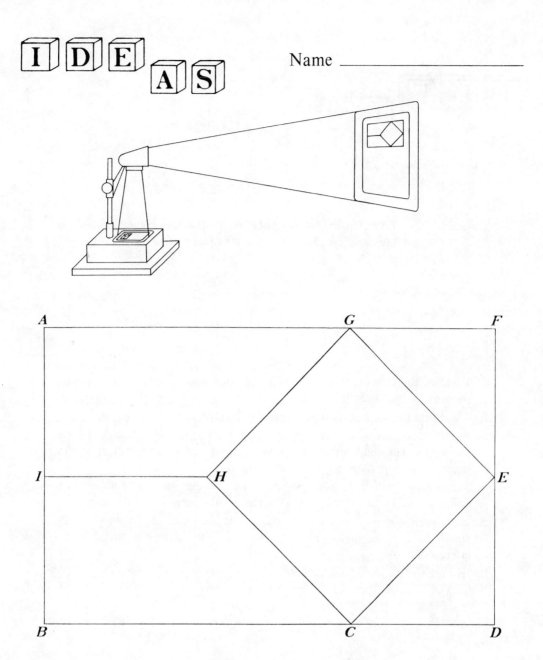

Measures of images

Length	Angle measure	Perimeter	Area
$\overline{DF}$_____	∠GFE_____	CEGH_____	ABDF_____
$\overline{FE}$_____	∠FGE_____	ABDF_____	CEGH_____
$\overline{GE}$_____	∠GHI_____	AGHI_____	
$\overline{FA}$_____			
$\overline{AG}$_____			

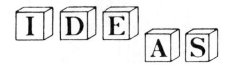

For Teachers

Objective: Experience in the investigation of a physical geometry situation that involves constant ratio and invariance of angle measure.

Levels: 6, 7, or 8

Directions for teachers:

1. Remove the activity sheet. Make a copy for each student and a transparency for your use.
2. Make sure that each student has a ruler and a protractor.
3. Before class, set your projector so that the image of segment $\overline{DF}$ on the screen (or chalkboard) is 14 inches long. (Avoid distortion of the figure by setting the projector on the same level as the screen.)
4. Have each student measure segment $\overline{DF}$ on his sheet. Then have a volunteer measure the image of $\overline{DF}$. Record the measure of the image.
5. Have each student complete the activity. Be sure that all students understand that only the measures of the image on the screen (or chalkboard) are recorded.
6. Discourage students from actually measuring the image lengths.
Note: You will need a chalkboard protractor.

Comments: The activity can be varied by changing the distance of the projector from the screen (or chalkboard). A discussion following the activity should bring out the patterns. The fact that the angle measure is constant will surprise many students.

Match a figure with each picture on the overhead projector.

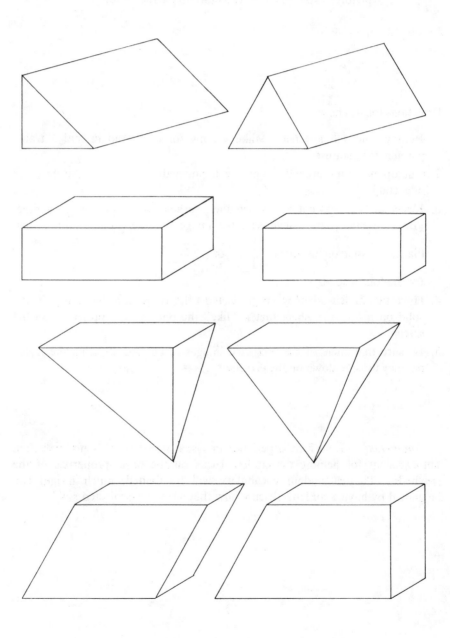

 For Teachers

Objectives: Experience in matching pictures of a geometric solid when the pictures show different orientations of the solid.

Levels: 2, 3, or 4

Directions for teachers:

1. Remove the activity sheet. Make a copy for each student and a transparency for your use.
2. Cut up the transparency so that each polyhedron is on a separate piece of acetate.
3. Place the "transparent solids" on the overhead projector one at a time. (Be sure that your student sees the image in different orientation than that of the solid on his paper: or   for the fourth solid.)
4. Have the student identify the projected solid by placing his finger on the solid on his activity sheet that is "like" the one he sees projected on the screen.
5. Be sure that some of the projected images result from flipping the transparency up-side-down on the projector.

Comments: This student experience in visual translations is not only fun, but excellent for helping the student focus on the basic properties of the polyhedra. Though student vocabulary will be limited, much insight can be gained by having students discuss how they view the projected solid.

IDEAS Name________________

How many rectangles?

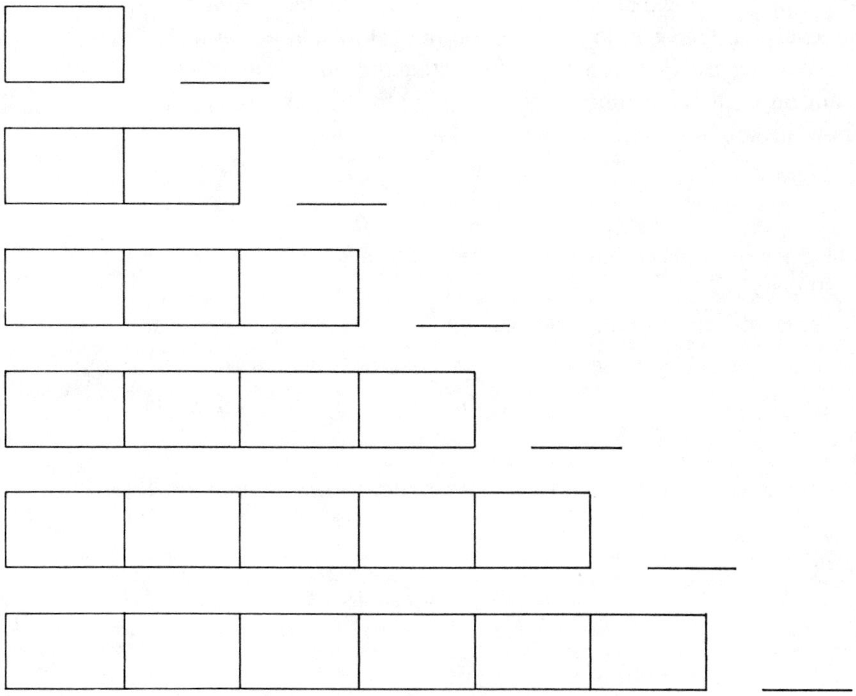

Complete the table.

Number of small rectangles	1	2	3	4	5	6	7	8	9	10
Total number of rectangles			6							

Objective: Experience with counting patterns

Level: 1, 2, 3, 4, 5, 6, 7, or 8

Comments: This set of experiences is nongraded. We have found that students from level 1 to level 8 enjoy working on the "How many rectangles?" activity sheet. We suggest that you start your students on the first activity sheet. Try them on each succeeding activity sheet until they experience frustration or show lack of interest.

Directions for teachers:

1. Remove the activity sheets that you decide to use and reproduce a copy for each student.

2. Start the students on counting the rectangles. Some students at every age level will have difficulty seeing the "composite" rectangles: and . Some will fail to count the "three-rectangle rectangle" .

3. Expect many students to be able to see the pattern in the table and to be able to extend their entries beyond those that were actually counted.

Answer keys and pattern clues:

IDEAS

Number of small (*n*)	1	2	3	4	5	6	7	8	9	10
Total number (*t*)	1	3	6	10	15	21	28	36	45	55

Pattern clues: The total number of rectangles (*t*) for any number of small rectangles (*n*) can be found by adding *n* to the total number for $n - 1$. (For 5 small rectangles, add 5 to the total of 4 small rectangles: $10 + 5 = 15$.)

Rule: $t = \dfrac{n(n + 1)}{2}$

Name _____

How many squares?

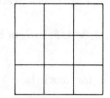

_____   _____   _____

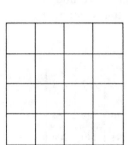

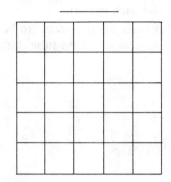

_____   _____

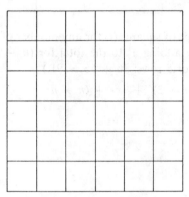

_____

Complete the table.

Number of rows	1	2	3	4	5	6	7	8
Number of small squares	1	4	9					
Total number of squares			14					

IDEAS

Objective: Experience with counting patterns

Level: 1, 2, 3, 4, 5, 6, 7, or 8

Directions for teachers:

1. Remove the activity sheets that you decide to use and reproduce a copy for each student.
2. Expect many students to be able to see the pattern in the table and to be able to extend their entries beyond those that were actually counted.

I D E A S

Rows (n)	1	2	3	4	5	6	7	8
Small	1	4	9	16	25	36	49	64
Total (t)	1	5	14	30	55	91	140	204

Pattern clues: The total number of squares (t) for any number of rows of small squares (n) can be found by adding n^2 to the total for $(n - 1)$ rows. (For 6 rows, add 36 to the total for 5 rows: $36 + 55 = 91$.)

Rule: $t = n^2 + (n - 1)^2 + (n - 2)^2 + \cdots + (n - n)^2$

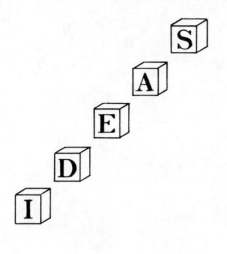

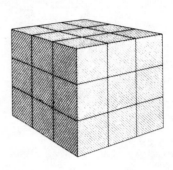

Name_____

Situation: Sugar cubes have been stacked to form a cube. The outside of the large cube has been colored with red food coloring.

1. How many sugar cubes were used?_____

2. How many sugar cubes are painted

    a. on four sides?_____

    b. on three sides?_____

    c. on two sides?_____

    d. on one side?_____

    e. on *zero* sides?_____

3. Imagine you stacked 64 sugar cubes in the same way and painted the outside. Answer questions 2a through 2e for this cube and write your answers below.

    a. _____ b. _____ c. _____ d. _____ e. _____

4. Can you stack 1,000,000 sugar cubes on the teacher's desk? Explain your answer.

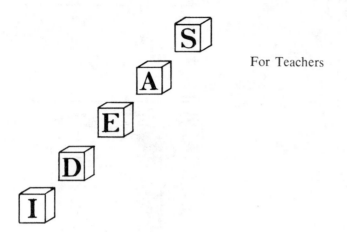

For Teachers

Objective: Experience in visualization of three dimensions

Grade level: 6, 7, or 8

Directions for teachers:

1. Remove the student worksheet and reproduce one copy for each student. (Option) Display single copy on bulletin board.
2. Provide a box of sugar cubes and food coloring to model the picture.

Directions for students:

1. Study the situation and answer the questions.
2. When you finish question 3, I want someone to make a model of a "64" cube for the class.
3. Question 4 is a tough one you can work on.

Comments: The questions asked about the pictured "27" cube are appropriate for everyone in your class. Few students will be able to answer the same questions for a "64" cube unless they either build a model or make a drawing of a 4-by-4-by-4 cube. A drawing can be made quite easily by tracing the cube shown and extending the edges as shown:

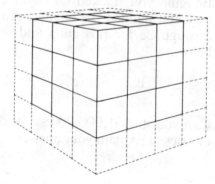

CORNER TO CORNER

| If you have a rectangle like this, 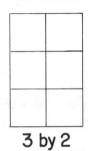 3 by 2 | and you draw a diagonal, 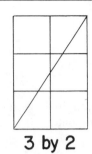 3 by 2 | the diagonal goes through 4 squares. 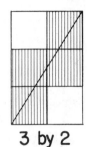 3 by 2 |

Investigate this diagonal pattern for other rectangles.

Use the squares below to draw rectangles. Use a straightedge.

	Dimensions	Squares cut with diagonal

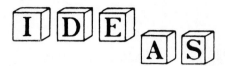

For Teachers

Objective: To investigate number patterns from a geometrical experiment

Levels: 3, 4, 5, 6

Directions for teachers:

1. Duplicate a worksheet for each child.
2. Make sure they understand the directions. It would be helpful to have additional squared paper available.
3. Organizing the data makes it easier to investigate the patterns. One suggestion for doing this is to look at all the rectangles that have one dimension kept the same; for example:

Dimensions	How many?
2 × 1	2
2 × 2	2
2 × 3	4
2 × 4	4
2 × 5	6

You may want to have different groups of students try the experiment keeping a different dimension constant.

THERE'S MORE THAN ONE WAY TO CUT
A CAKE

| If you cut across a cake with 1 straight cut, you'll have 2 pieces | If you use 2 straight cuts, you can do it so you'll have either 3 pieces .... or 4 pieces |

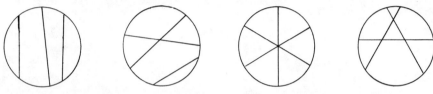

With 3 straight cuts, there are 4 different ways

*Remember, every cut must be straight and across the cake

Investigate this cake cutting pattern. Record what you find on the table below. Continue the table on another paper.

How many cuts?	Drawing	Pieces	How many ways?
1		2	} 1
2		3 4	} 2
3		4 5 6 7	} 4
4			

Can you predict what is the greatest number of pieces you could get from 10 cuts? Try it and see!

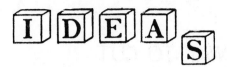 For Teachers

Levels: 2, 3, 4

Objective: To investigate patterns in a geometrical partitioning experiment

Directions for teachers:

1. Duplicate a worksheet for each child.

2. Make sure they understand the directions.

3. When all the children have had the chance to investigate this problem, discuss with the entire class the patterns they found.

Name _____

There's more than one way to ✂ CUT a SQUARE !

Here's a square cut into 4 smaller squares. ☞

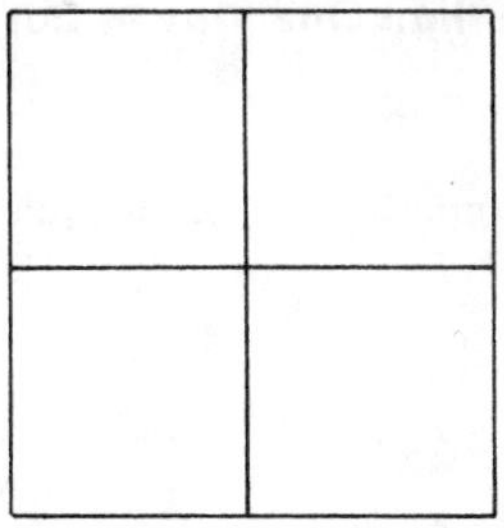

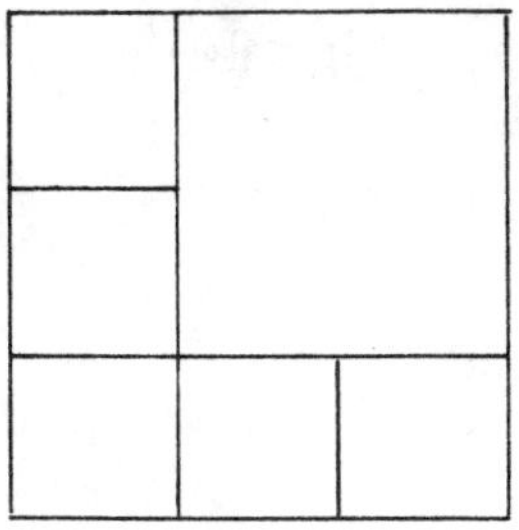

☞ This square has been cut into 6 smaller squares.

Find ways to cut squares into the number of smaller squares indicated below.

7

8

9

Draw more squares and keep on going.

10

11

12

13

14

15

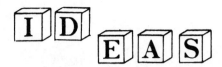

For Teachers

There's More than One Way to Cut a Square

Levels 4, 5, 6, 7, 8

Here are some ways to cut a square.

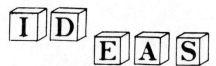

Find each line path from START to FINISH.

Measure the length of each path in centimeters.

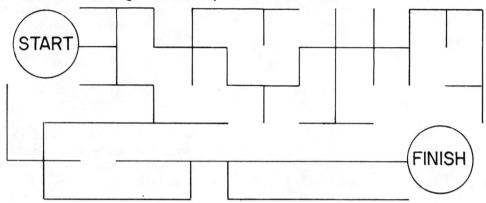

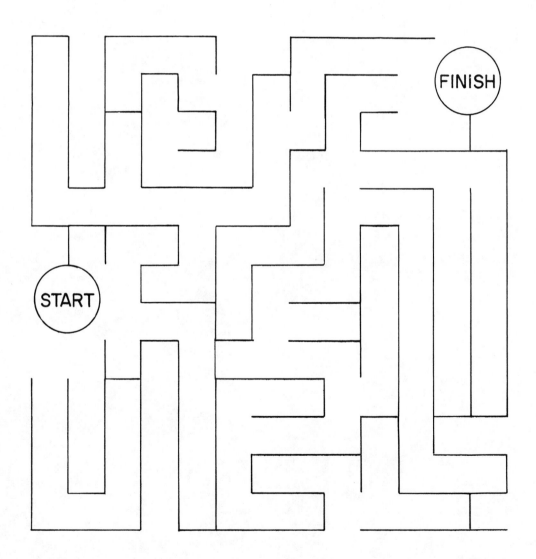

 For Teachers

Objective: To practice measuring line segments

Level: 4, 5, 6

Directions for teachers:

1. Give each student a copy of the worksheet.
2. Be sure each student has a centimeter ruler.
3. Let the students read the directions and go to work. The students are expected to measure each straight section of the path to the nearest centimeter and then to add these measurements.

Going further:

1. Ask students to put an "X" on each part of the maze that is 6 cm (or 5 cm or 4 cm) long.
2. Let each student make a maze and have another student find and measure the path.

Answers: first maze, 20 cm; second maze, 47 cm.

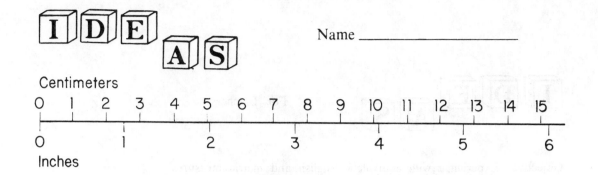

Name _____

Put an X on the line for each measure.

2 inches   3½ inches   5 inches

10 centimeters   5 centimeters   1 centimeter

Use the number line to approximate the metric measure for each English measure.

2 inches, ____ centimeters   4 inches, ____ centimeters

5½ inches, ____ centimeters

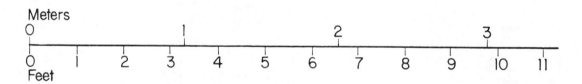

Put an X on the line for each measure. (Remember: 100 cm. = 1 meter.)

3 feet   2 meters   6½ feet

150 centimeters   9 feet 10 inches   2½ meters

Use the number line to approximate the metric measure for each English measure.

3¼ feet, ____ meters   8 feet, ____ meters

5 feet, ____ meters

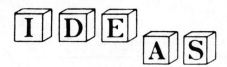 For Teachers

Objective: Experience with equivalent English and metric measures

Levels: 4, 5, 6, 7, or 8

Directions for teachers:

1. Remove the activity sheets that you feel are appropriate for your students. Make copies for each student.

2. Use one or more of these activity sheets in a measurement sequence. Note that the names of some units are abbreviated in the drawings on these sheets.

3. If students disagree regarding certain answers, they should be encouraged to perform a physical measurement to determine the correct answer.

Comments: These activities are not meant to replace the laboratory experiences that build the student's basic referent for measurement and the various units involved. Their use should follow laboratory experiences that include actual measuring of length, weight, and volume using instruments that measure in metric units. Contemporary science programs include the necessary measuring instruments calibrated in metric units. If the instruments are not available in your school, contact your science supervisor.

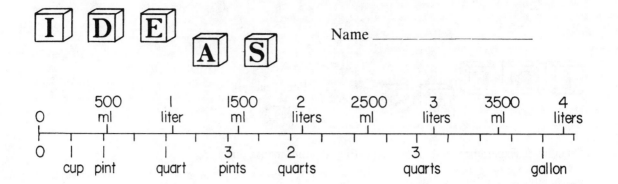

Name _____

Put an X on the measurement line for each measure.

1 cup	3 cups	2 pints
$1\frac{1}{2}$ liters	$2\frac{1}{2}$ quarts	8 cups
2.8 liters	495 milliliters	3550 milliliters

Use the measurement line to estimate the metric measure
for each English measure.

Use the measurement line to estimate the English measure
for each metric measure.

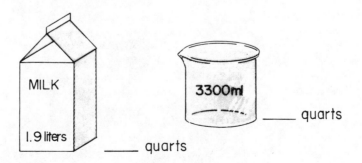

_____ quarts   _____ quarts

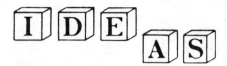 For Teachers

Objective: Experience with equivalent English and metric measures

Levels: 4, 5, 6, 7, or 8

Directions for teachers:

1. Remove the activity sheets that you feel are appropriate for your students. Make copies for each student.
2. Use one or more of these activity sheets in a measurement sequence. Note that the names of some units are abbreviated in the drawings on these sheets.
3. If students disagree regarding certain answers, they should be encouraged to perform a physical measurement to determine the correct answer.

Comments: These activities are not meant to replace the laboratory experiences that build the student's basic referent for measurement and the various units involved. Their use should follow laboratory experiences that include actual measuring of length, weight, and volume using instruments that measure in metric units. Contemporary science programs include the necessary measuring instruments calibrated in metric units. If the instruments are not available in your school, contact your science supervisor.

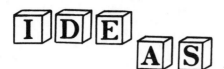

Name _____

Metric

0   100g 200g 300g 400g 500g 600g 700g 800g 900g 1kg   1100g

0   4oz.   8oz. 12oz. 1lb.   4oz.   8oz. 12oz. 2lbs 4oz.   8oz.

English

Put an X on the measurement line for each measure.

8 ounces	9 ounces	$\frac{1}{2}$ pound
1 pound 2 ounces	$1\frac{1}{4}$ pound	2.2 pounds
40 grams	100 grams	500 grams
10 grams	1000 grams	1050 grams

Estimate the weight in grams of each object.

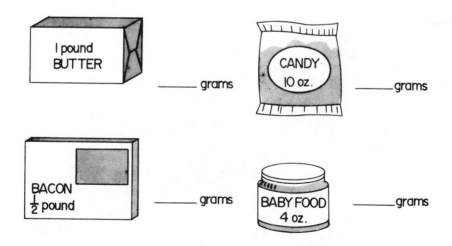

1 pound BUTTER   _____ grams

CANDY 10 oz.   _____ grams

BACON $\frac{1}{2}$ pound   _____ grams

BABY FOOD 4 oz.   _____ grams

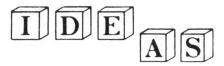 For Teachers

Objective: Experience with equivalent English and metric measures

Levels: 4, 5, 6, 7, or 8

Directions for teachers:

1. Remove the activity sheets that you feel are appropriate for your students. Make copies for each student.
2. Use one or more of these activity sheets in a measurement sequence. Note that the names of some units are abbreviated in the drawings on these sheets.
3. If students disagree regarding certain answers, they should be encouraged to perform a physical measurement to determine the correct answer.

Comments: These activities are not meant to replace the laboratory experiences that build the student's basic referent for measurement and the various units involved. Their use should follow laboratory experiences that include actual measuring of length, weight, and volume using instruments that measure in metric units. Contemporary science programs include the necessary measuring instruments calibrated in metric units. If the instruments are not available in your school, contact your science supervisor.

Name_____

m means meter

cm means centimeter

1 m = 100 cm

Circle the best measurement for each object pictured below.

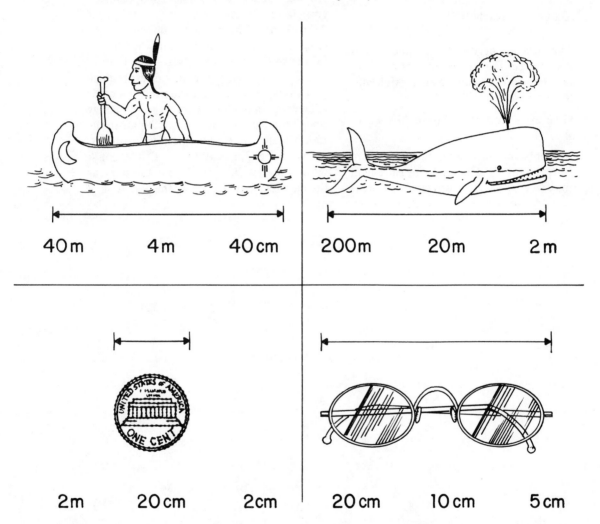

40 m   4 m   40 cm   200 m   20 m   2 m

2 m   20 cm   2 cm   20 cm   10 cm   5 cm

Circle the measurement in each pair that represents the shorter length.

4 cm, 40 m   8 m, 800 cm

7000 cm, 7 m   10 m, 100 cm

 For Teachers

Objective: To estimate measurements and to use the relationship between meter and centimeter

Level: 6, 7, 8

Directions for teachers:

1. The students need to have some experience measuring in centimeters and meters before they can be successful with the exercises.

2. Give each student a copy of the worksheet.

3. Students should imagine the actual objects pictured. The best measurement may not be exactly correct, so students should choose the one which is closest.

Ask students to explain how they decided which measurement is best.

4. Ask students to make up exercises like those on the worksheet.

Answers: canoe, 4 m; whale, 20 m; penny, 2 cm; glasses, 10 cm;
4 cm, 8 m = 800 cm, 7 m, 100 cm.

Comments: There are several ways to obtain answers to the exercises at the bottom of the page. In the first pair, 4 is less than 40 and cm is a smaller unit than m, so 4 cm is smaller than 40 m. Alternatively change one of the measurements in each pair to the unit of the other measurement; for example, since 7 m = 700 cm and 7000 cm = 70 m, the pair 7000 cm, 7 m is the same as the pair 7000 cm, 700 cm or the pair 70 m, 7 m.

I D E A S

CHALK

Length: 8 centimeters
Weight: 10 grams

Estimate the total weight and the total length of the
pieces of chalk shown for each exercise.

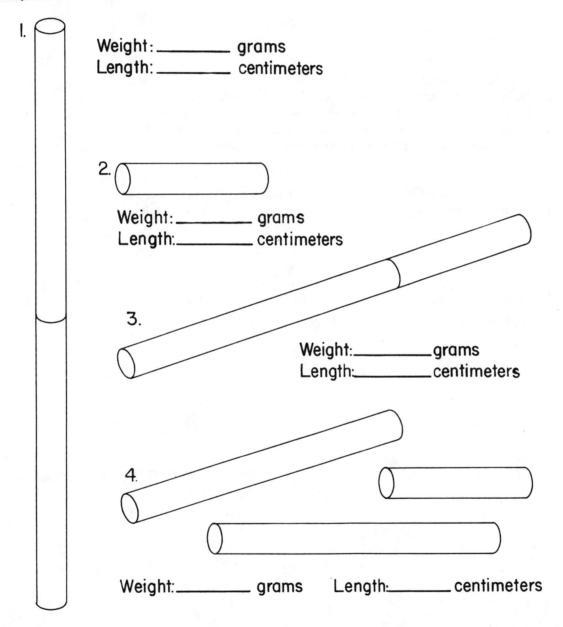

1. Weight: _____ grams
   Length: _____ centimeters

2. Weight: _____ grams
   Length: _____ centimeters

3. Weight: _____ grams
   Length: _____ centimeters

4. Weight: _____ grams   Length: _____ centimeters

 For Teachers

Objecti . Experience with weight and length using the metric system

Gra   level: 3 or 4

Directions for teachers:

Reproduce a copy of the activity sheet for each student. You may wish to exhibit or even pass around several new pieces of chalk. Students may note that most pieces of chalk are not identical and correctly conclude that the 8 centimeters and 10 grams are approximations. Once the students have the basic referent in mind they should work independently on this activity.

Comments: The total length as an end-to-end chain of individual pieces is an important concept. It should not be expected that this concept is intuitive for all students.

Answers

1. 20 grams, 16 centimeters   2.   5 grams, 4 centimeters
3. 15 grams, 12 centimeters   4. 25 grams, 20 centimeters

Name _____

Measure each body part to the nearest centimeter.

height

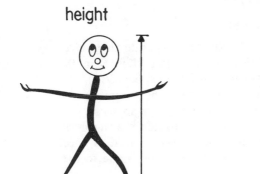

_____cm

head

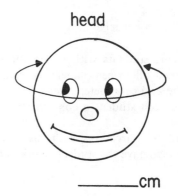

_____cm

arm span

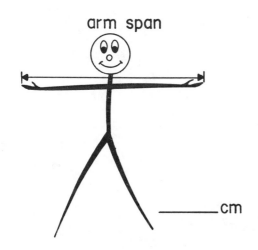

_____cm

smile

_____cm

knee

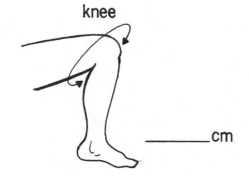

_____cm

big toe

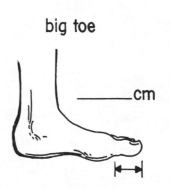

_____cm

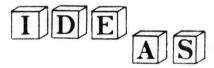

For Teachers

Objective: To develop a frame of reference for centimeter by measuring parts of the body

Level: 2, 3, 4, 5

Directions for teachers:

1. Provide students with measuring tapes marked off in centimeters, or pieces of string and centimeter rulers.
2. Give each student a copy of the worksheet.
3. Read the directions with the students and then have the students work in pairs or small groups to measure each other's body parts.

Going further:

1. Have the students separate themselves into three groups: (see IDEAS in the *Arithmetic Teacher,* October 1974)
   a. tall rectangles (height > arm span)
   b. squares (height = arm span)
   c. short rectangles (height < arm span)
2. Find approximate ratios: head to knee, knee to toe, head to toe, smile to toe, and so on.

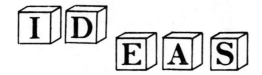

BODY RATIOS

Compare your body measures.

Record like this. ————————⟶
        Foot to height = 6 : 1

Around wrist to height = _____

Around wrist to waist   = _____

Around head to height = _____

Around fist to waist = _____

Around wrist to neck = _____

Around neck to arm = _____

Arm to height = _____

Around fist to foot = _____

Foot to arm = _____

Around neck to height = _____

Compare your ratios with a friend's ratios.

 For Teachers

Objective: Investigating ratios using body measures.

Levels: 6, 7, or 8

Directions for teachers:

1. Remove the activity sheet BODY RATIOS and reproduce one copy for each student.

2. Give each student a piece of string. Make sure it doesn't have any stretch and is longer than he or she is tall.

3. After all have finished, have the students compare similarities and differences.

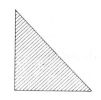

How many of these triangular units would it take to make each polygon?

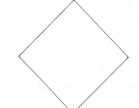

A. _____

B. _____

C. _____

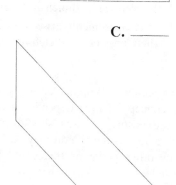

D. _____

E. _____

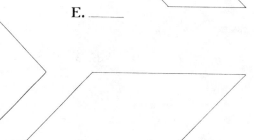

F. _____

G. _____

For Teachers

Objective: Experience with the area concept using a standard unit that is not a square

Grade level: 6, 7, or 8

Directions for teachers:

1. Remove the activity sheet and reproduce a copy for each student.
2. Have the students work independently. Cut the corners off a file card so that each corner is congruent to the triangle pictured at the top of the activity sheet. "Loan" one or more of these to the least spatially oriented students to help them get started.
3. Encourage as much independence as possible.
4. If disagreements arise in checking their answers, have students "prove" their answer by sketching the triangles that fill the polygon in question.

Comments: Whether or not you choose to formally relate this activity to the concept of area depends on the student's previous exposure to area. His first exposure to area should probably involve tessellations of rectangles with a unit square. You may wish to encourage creativity by having students draw other polygons that could be formed by specific numbers of the unit triangle.

Answers

A. 2   *B.* 2   *C.* 4   *D.* 6   *E.* 3   *F.* 8   *G.* 4

I D E A S

THE
IMAGINARY
GARDEN AND FENCE PROBLEM

You are planning a garden that will cover 12 squares on the squared paper.

Mathematicians would say:
"The area of the garden
is 12 square centimeters."

You may make the garden
any shape you like.

But....you need to build
a fence around it too.

Draw different possible gardens on the squared paper and find which needs the shortest fence and which needs the longest

Mathematicians would say:
Find which shape of 12 sq.
cm has the shortest perimeter
and which has the longest.

Cut out your results and paste them here.

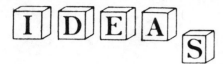 For Teachers

Objective: To develop the idea of the relationship between perimeter and area of rectangles

Levels: 4, 5, and 6

Directions for teachers:

1. Reproduce a copy of the activity sheet for each student.
2. Provide squared paper.
3. The student directions are self-explanatory.

Going further:

When the students have solved the 12-square problems, ask them to solve problems using 16, 25, or 36 squares.

Name _____

SHAPES THAT GROW

	Area of your shape	Area after you doubled the sides

1. Draw a shape on the centimeter squared paper. Count the squares inside the shape and write the number here — (This is the area in sq. cm)

2. Now draw the same shape, but make each side twice as long. Find the area of this larger shape and write it here —

3. Do this for at least 5 different shapes. Continue recording on the chart.

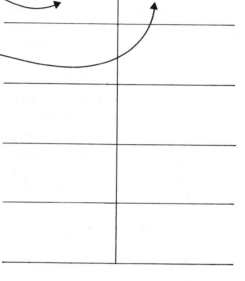

Example:

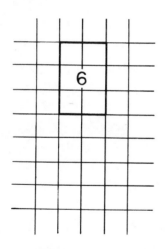

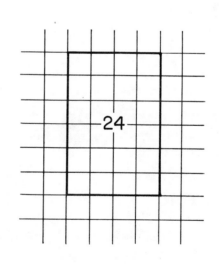

6	24

What pattern do you notice? _____

_____

_____

_____

_____

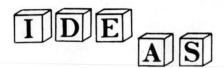

For Teachers

Objective: Experience with the relationship between perimeter and area

Levels: 4, 5, or 6

Directions for teachers:

1. Remove the activity sheet and reproduce a copy for each student.

2. Students will need centimeter squared paper for this activity.

3. Make sure your students understand the directions before they start.